Carbon Capture and Storage

Carbon Capture and Storage

Madison Reed
Trevor Boyd

Kruger Brentt
Publishers
2026

Kruger Brentt Publishers UK. LTD.
Company Number 9728962

Regd. Office: 68 St Margarets Road, Edgware, Middlesex HA8 9UU

ISBN: 978-1-78715-605-0

For information on all our publications visit our website at http://krugerbrentt.com/

Preface

Carbon Capture and Storage explores one of the most promising technological approaches to mitigating climate change: the capture, transport, and storage of carbon dioxide (CO_2) emissions from industrial and energy-related sources. As global concerns over rising greenhouse gas concentrations intensify, carbon capture and storage (CCS) has emerged as a critical strategy for reducing the impact of fossil fuel use while transitioning to a low-carbon economy. This book aims to provide a thorough introduction to the principles, technologies, and practical applications of CCS for students, researchers, and professionals in environmental science, engineering, and energy sectors.

The book begins with an overview of climate change, the role of CO_2 emissions, and the need for carbon management. It then introduces the fundamental concepts of CCS, including the different methods of capturing CO_2—post-combustion, pre-combustion, and oxy-fuel combustion—as well as transport options via pipelines and storage techniques such as geological sequestration, mineralization, and ocean storage. The technical, economic, and environmental considerations of each method are discussed in detail, emphasizing both opportunities and challenges.

A key focus of the text is on the design, implementation, and monitoring of CCS projects. It examines the engineering aspects of capture systems, storage site selection, risk assessment, and long-term monitoring to ensure the safety and effectiveness of CO_2 sequestration. Additionally, the book explores policy frameworks, regulatory challenges, and the role of international cooperation in promoting CCS adoption as part of broader climate mitigation strategies.

Throughout the book, case studies and real-world examples illustrate successful CCS projects, highlighting lessons learned, technological innovations, and strategies for overcoming practical challenges. Emphasis is placed on the integration of CCS with renewable energy systems and other climate solutions to create sustainable pathways toward net-zero emissions.

We hope that Carbon Capture and Storage will serve as a valuable resource for students, researchers, and industry practitioners seeking to understand and contribute to the development and deployment of CCS technologies. By providing both theoretical foundations and practical insights, this book aims to support the global effort to mitigate climate change and advance a sustainable energy future.

Madison Reed | Trevor Boyd

Contents

CHAPTER 1

Introduction to Carbon Capture and Storage (CCS)

1.1 WHAT IS CARBON CAPTURE AND STORAGE (CCS)?

Carbon capture and storage (CCS) is a process by which carbon dioxide (CO_2) from industrial installations is separated before it is released into the atmosphere, then transported to a long-term storage location. The CO_2 is captured from a large point source, such as a natural gas processing plant and is typically stored in a deep geological formation. Around 80% of the CO_2 captured annually is used for enhanced oil recovery (EOR), a process by which CO_2 is injected into partially depleted oil reservoirs in order to extract more oil and then is largely left underground. Since EOR utilizes the CO_2 in addition to storing it, CCS is also known as carbon capture, utilization, and storage (CCUS).

Oil and gas companies first used the processes involved in CCS in the mid 20th century. Early CCS technologies were mainly used to purify natural gas and increase oil production. Beginning in the 1980s and accelerating in the 2000s, CCS was discussed as a strategy to reduce greenhouse gas emissions. Around 70% of announced CCS projects have not materialized, with a failure rate above 98% in the electricity sector. As of 2024 CCS was in operation at 44 plants worldwide, collectively capturing about one-thousandth of global carbon dioxide emissions. 90% of CCS operations involve the oil and gas industry. Plants with CCS require more energy to operate, thus they typically burn additional fossil fuels and increase the pollution caused by extracting and transporting fuel.

CCS could have a critical but limited role in reducing greenhouse gas emissions. However, other emission-reduction options such as solar and wind energy, electrification, and public transit are less expensive than CCS and are much more effective at reducing air pollution. Given its cost and limitations, CCS is envisioned to be most useful in specific niches. These niches include heavy industry and plant retrofits. In the context of deep and sustained cuts in natural gas consumption, CCS can reduce emissions from natural gas processing. In electricity generation and hydrogen production, CCS is envisioned to complement a broader shift to renewable energy. CCS is a component of bioenergy with carbon capture and storage, which can under some conditions remove carbon from the atmosphere.

The effectiveness of CCS in reducing carbon emissions depends on the plant's capture efficiency, the additional energy used for CCS itself, leakage, and business and technical issues that can keep facilities from operating as designed. Some large CCS implementations have sequestered far less CO_2 than originally expected. Controversy remains over whether using captured CO_2 to extract more oil ultimately benefits the climate. Many environmental groups regard CCS as an unproven, expensive technology that perpetuates fossil fuel dependence. They believe other ways to reduce emissions are more effective and that CCS is a distraction.

Some international climate agreements refer to the concept of fossil fuel abatement, which is not defined in these agreements but is generally understood to mean use of CCS. Almost all CCS projects operating today have benefited from government financial support. Countries with programs to support or mandate CCS technologies include the US, Canada, Denmark, China, and the UK.

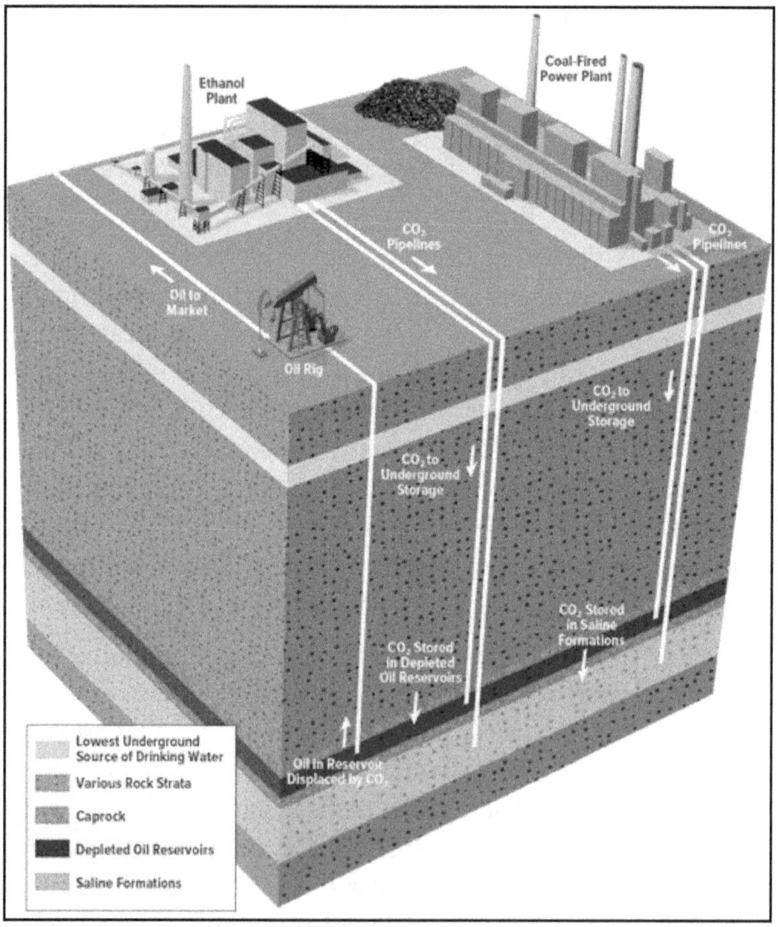

Fig: With CCS, Carbon Dioxide is Captured from a Point Source, such as an Ethanol Refinery. It is Usually Transported via Pipelines and then Either Used to Extract Oil or Stored in a Dedicated Geologic Formation

1.1.1 History and Current Status

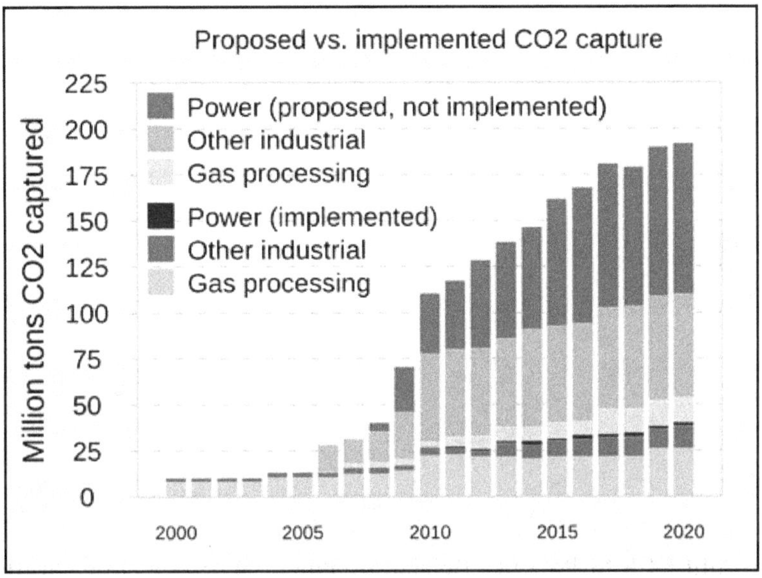

Fig: Global Proposed (Grey Bars) vs. Implemented (Blue Bars) Annual CO_2 Captured. Both are in Million Tons of CO_2 Per Annum (Mtpa). More than 75% of Proposed CCS Installations for Natural-Gas Processing have been Implemented

In the natural gas industry, technology to remove CO_2 from raw natural gas was patented in 1930. This processing is essential to make natural gas ready for commercial sale and distribution. Usually after CO_2 is removed, it is vented to the atmosphere. In 1972, American oil companies discovered that CO_2 could profitably be used for EOR. Subsequently, natural gas companies in Texas began capturing the CO_2 produced by their processing plants and selling it to local oil producers for EOR.

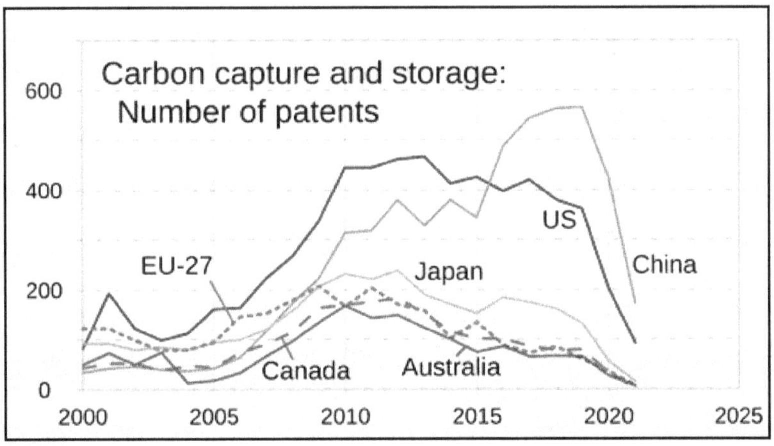

The Number of Patents Covering CCS Technologies Surged in the 2000s, but Stalled or Declined in the 2010s

Fig: Plans to Add CCS to Poland's Bełchatów Power Station were Cancelled in 2013. Over 98% of Plans to use CCS in Power Plants have Failed

The use of CCS as a means of reducing human-caused CO_2 emissions is more recent. In 1977, the Italian physicist Cesare Marchetti proposed that CCS could be used to reduce emissions from coal power plants and fuel refineries. Small-scale implementations were first demonstrated in the early 1980s and an economic evaluation was published in 1991. The first large-scale CO_2 capture and injection project with dedicated CO_2 storage and monitoring was commissioned at the Sleipner gas field in Norway in 1996.

In 2005, the IPCC released a report highlighting CCS, leading to increased government support for CCS in several countries. Governments spent an estimated USD $30 billion on subsidies for CCS and for fossil-fuel-based hydrogen. Globally, 149 projects to store 130 million tonnes of CO_2 annually were proposed to be operational by 2020. Of these, around 70% were not implemented. Limited one-off capital grants, the absence of measures to address long-term liability for stored CO_2, high operating costs, limited social acceptability and vulnerability of funding programmes to external budget pressures all contributed to project cancellations.

In 2020, the International Energy Agency (IEA) stated, "The story of CCUS has largely been one of unmet expectations: its potential to mitigate climate change has been recognised for decades, but deployment has been slow and so has had only a limited impact on global CO_2 emissions."

By July 2024, commercial-scale CCS was in operation at 44 plants worldwide. Sixteen of these facilities were devoted to separating naturally-occurring CO_2 from raw natural gas. Seven facilities were for hydrogen, ammonia, or fertilizer production, seven for chemical production, five for electricity and heat, and two for oil refining. CCS was also used in one iron and steel plant. Additionally, three facilities worldwide were devoted to CO_2 transport/storage. As of 2024, the oil and gas industry is involved in 90% of CCS capacity in operation around the world. Collectively, the facilities capture about one-thousandth of global greenhouse gas emissions.

Eighteen facilities were in the United States, fourteen in China, five in Canada, and two in Norway. Australia, Brazil, Qatar, Saudi Arabia, and the United Arab Emirates had one project each. As of 2020, North America has more than 8,000 km (5,000 mi) of CO_2 pipelines, and there are two CO_2 pipeline systems in Europe and two in the Middle East.

1.1.2 Process Overview

CCS facilities capture carbon dioxide before it enters the atmosphere. Generally, a chemical solvent or a porous solid material is used to separate the CO_2 from other components of a plant's exhaust stream. Most commonly, the gas stream passes through an amine solvent, which binds the CO_2 molecule. This CO_2-rich solvent is heated in a regeneration unit to release the CO_2 from the solvent. The CO_2 stream then undergoes conditioning to remove impurities and bring the gas to an appropriate temperature for compression. The purified CO_2 stream is compressed and transported for storage or end-use and the released solvents are recycled to capture more CO_2 from the facility. After the CO_2 has been captured, it is usually compressed into a supercritical fluid and then injected underground. Pipelines are the cheapest way of transporting CO_2 in large quantities onshore and, depending on the distance and volumes, offshore. Transport via ship has been researched. CO_2 can also be transported by truck or rail, albeit at higher cost per tonne of CO_2.

1.1.3 Technical Components

CCS processes involve several different technologies working together. Technological components are used to separate and treat CO_2 from a gas mixture, compress and transport the CO_2, inject it into the subsurface, and monitor the overall process. There are three ways that CO_2 can be separated from a gas mixture: post-combustion capture, pre-combustion capture, and oxy-combustion:

- In post combustion capture, the CO_2 is removed after combustion of the fossil fuel.
- The technology for pre-combustion is widely applied in natural gas processing. In these cases, the fossil fuel is partially oxidized, for instance in a gasifier. The CO from the resulting syngas (CO and H2) reacts with added steam (H2O) and is shifted into CO_2 and H2. The resulting CO_2 can be captured from a relatively pure exhaust stream.

The H2 can be used as fuel. Several advantages and disadvantages apply versus post combustion capture.

- In oxy-fuel combustion the fuel is burned in pure oxygen instead of air. The gas that is released consists of mostly CO_2 and water vapor. After water vapor is condensed through cooling, the result is an almost pure CO_2 stream. A disadvantage of this technique is that it requires a relatively large amount of oxygen, which is expensive and energy-intensive to produce.

Absorption, or carbon scrubbing with amines is the dominant capture technology. Other technologies proposed for carbon capture are membrane gas separation, chemical looping combustion, calcium looping, and use of metal-organic frameworks and other solid sorbents. Impurities in CO_2 streams, like sulfur dioxides and water vapor, can have a significant effect on their phase behavior and could cause increased pipeline and well corrosion. In instances where CO_2 impurities exist, a process is needed to remove them.

1.1.4 Storage and Enhanced Oil Recovery

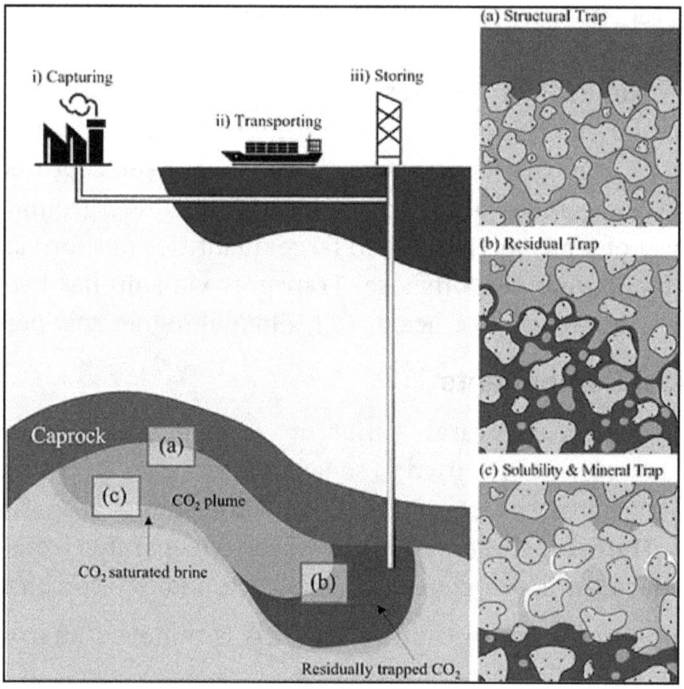

Fig: Diagram of Mechanisms for Trapping Carbon Dioxide in Dedicated Geologic Storage

Storing CO_2 involves the injection of captured CO_2 into a deep underground geological reservoir of porous rock overlaid by an impermeable layer of rocks, which seals the reservoir and prevents the upward migration of CO_2 and escape into the atmosphere. The gas is usually compressed first into a supercritical fluid. When the compressed CO_2 is injected into a reservoir, it flows through it, filling the pore space. The reservoir must be at depths greater than 800 m (2,600 ft) to retain the CO_2 in a fluid state. As of 2024, around 80% of the CO_2 captured annually is used for enhanced oil recovery (EOR). In EOR, CO_2 is injected into partially depleted oil fields to enhance production. The CO_2 binds with oil to make it less dense, allowing oil to rise to the surface faster. The addition of CO_2 also increases the overall reservoir pressure, thereby improving the mobility of the oil, resulting in a higher flow of oil towards the production wells. Depending on the location, EOR results in around two additional barrels of oil for every tonne of CO_2 injected into the ground and using that oil produces approximately one tonne of CO_2. Oil extracted through EOR is mixed with CO_2, which can then mostly be recaptured and re-injected multiple times. This CO_2 recycling process can reduce losses to 1%; however, it is energy-intensive.

Around 20% of captured CO_2 is injected into dedicated geological storage, usually deep saline aquifers. These are layers of porous and permeable rocks saturated with salty water. Worldwide, saline formations have higher potential storage capacity than depleted oil wells. Dedicated geologic storage is generally less expensive than EOR because it does not require a high level of CO_2 purity and because suitable sites are more numerous, which means pipelines can be shorter.

Various other types of reservoirs for storing captured CO_2 were being researched or piloted as of 2021: CO_2 could be injected into coal beds for enhanced coal bed methane recovery. Ex-situ mineral carbonation involves reacting CO_2 with mine tailings or alkaline industrial waste to form stable minerals such as calcium carbonate. In-situ mineral carbonation involves injecting CO_2 and water into underground formations that are rich in highly-reactive rocks such as basalt. There, the CO_2 may react with the rock to form stable carbonate minerals relatively quickly. Once this process is complete, the risk of CO_2 escape from carbonate minerals is estimated to be close to zero. The global capacity for underground CO_2 storage is potentially very large and is unlikely to be a constraint on the development of CCS. Total storage capacity has been estimated at between 8,000 and 55,000 gigatonnes. However, a smaller fraction will most likely prove to be technically or commercially feasible. Global capacity estimates are uncertain, particularly for saline aquifers where more site characterization and exploration is still needed.

Long-Term CO_2 Leakage

In geologic storage, the CO_2 is held within the reservoir through several trapping mechanisms: structural trapping by an impermeable rock layer called a caprock, solubility trapping in pore space water, residual trapping in individual or groups of pores, and mineral trapping by reacting with the reservoir rocks to form carbonate minerals. Mineral trapping progresses over time but is extremely slow. After injection, supercritical CO_2 tends to rise until it is trapped beneath a caprock. Once it encounters a caprock, it spreads laterally until it encounters a gap. If there are fault planes near the injection zone, CO_2 could migrate along the fault to the surface, leaking into the atmosphere, which would be potentially dangerous to life in the surrounding area.

If the injection of CO_2 creates pressures underground that are too high, the formation will fracture, potentially causing an earthquake. While research suggests that earthquakes from injected CO_2 would be too small to endanger property, they could be large enough to cause a leak.

According to the IPCC, well-managed storage sites likely retain over 99% of injected CO_2 for more than a thousand years, where 'likely' means a 66–90% probability. Estimates of long-term leakage rates rely on complex simulations since field data is limited. If very large amounts of CO_2 are sequestered, even a 1% leakage rate over 1000 years could cause significant impact on the climate for future generations.

1.1.5 Social and Environmental Impacts

Energy and Water Requirements

Facilities with CCS use more energy than those without CCS. The energy consumed by CCS is called an "energy penalty". The energy penalty of CCS varies depending on the source of CO_2. If the gas from the source has a very high concentration of CO_2, additional energy is needed only to dehydrate, compress, and pump the CO_2. If the facility produces gas with a lower concentration of CO_2, as is the case for power plants, energy is also required to separate CO_2 from other gas components.

Early studies indicated that to produce the same amount of electricity, a coal power plant would need to burn 14–40% more coal and a natural gas combined cycle power plant would need to burn 11–22% more gas. When CCS is used in coal power plants, it has been estimated that about 60% of the energy penalty originates from the capture process, 30% comes from compression of the extracted CO_2, and the remaining 10% comes from pumps and fans. Depending on the technology used, CCS can require large amounts of water. For instance, coal-fired power plants with CCS may need to use 50% more water.

Pollution

Since plants with CCS require more fuel to produce the same amount of electricity or heat, the use of CCS increases the "upstream" environmental problems of fossil fuels. Upstream impacts include pollution caused by coal mining, emissions from the fuel used to transport coal and gas, emissions from gas flaring, and fugitive methane emissions. Since CCS facilities require more fossil fuel to be burned,

CCS can cause a net increase in air pollution from those facilities. This can be mitigated by pollution control equipment, however no equipment can eliminate all pollutants. Since liquid amine solutions are used to capture CO_2 in many CCS systems, these types of chemicals can also be released as air pollutants if not adequately controlled. Among the chemicals of concern are volatile nitrosamines and nitramines which are carcinogenic when inhaled or drunk in water.

Studies that consider both upstream and downstream impacts indicate that adding CCS to power plants increases overall negative impacts on human health. The health impacts of adding CCS in the industrial sector are less well-understood. Health impacts vary significantly depending on the fuel used and the capture technology.

After CO_2 injected into underground geologic formations, there is a risk of nearby shallow groundwater becoming contaminated. Contamination can occur either from movement of the CO_2 into groundwater or from movement of displaced brine. Careful site selection and long-term monitoring are necessary to mitigate this risk.

Sudden CO_2 Leakage

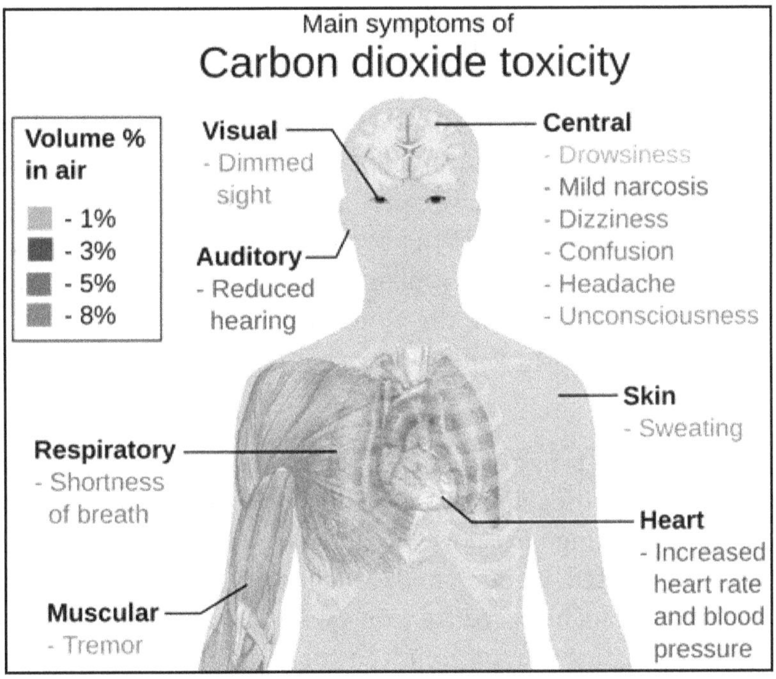

Fig: Main Symptoms of Carbon Dioxide Toxicity

CO_2 is a colorless and odorless gas that accumulates near the ground because it is heavier than air. In humans, exposure to CO_2 at concentrations greater than 5% (50,000 parts per million) causes the development of hypercapnia and respiratory acidosis. Concentrations of more than 10% may cause convulsions, coma, and death. CO_2 levels of more than 30% act rapidly leading to loss of consciousness in seconds.

Pipelines and storage sites can be sources of large accidental releases of CO_2 that can endanger local communities. A 2005 IPCC report stated that "existing CO_2 pipelines, mostly in areas of low population density, accident numbers reported per kilometre of pipeline are very low and are comparable to those for hydrocarbon pipelines." The report also stated that the local health and safety risks of geologic CO_2 storage were "comparable" to the risks of underground storage of natural gas if good site selection processes, regulatory oversight, monitoring, and incident remediation plans are in place. As of 2020, the ways that pipelines can fail is less well-understood for CO_2 pipelines than for natural gas or oil pipelines, and few safety standards exist that are specific to CO_2 pipelines.

While infrequent, accidents can be serious. In 2020 a CO_2 pipeline ruptured following a mudslide near Satartia, Mississippi, causing people nearby to lose consciousness. About 200 people were evacuated and 45 were hospitalized, and some experienced longer-term effects on their health. High concentrations of CO_2 in the air also caused vehicle engines to stop running, hampering the rescue effort.

Jobs

Retrofitting facilities with CCS can help to preserve jobs and economic prosperity in regions that rely on emissions-intensive industry, while avoiding the economic and social disruption of early retirements. For instance, Germany's plans to retire around 40 GW of coal-fired generation capacity before 2038 is accompanied by a EUR 40 billion (USD 45 billion) package to compensate the owners of coal mines and power plants as well as support the communities that will be affected. There is potential for reducing these costs if plants are retrofitted with CCS. Retrofitting CO_2 capture equipment can enable the continued operation of existing plants, as well as associated infrastructure and supply chains.

Equity

In the United States, the types of facilities that could be retrofitted with CCS are often located in communities that have already borne the negative environmental and health impacts of living near power or industrial facilities. These facilities are disproportionately located in poor and/or minority communities. While there is evidence that CCS can help reduce non-CO_2 pollutants along with capturing CO_2, environmental justice groups are often concerned that CCS will be used as a way to prolong a facility's lifetime and continue the local harms it causes. Often, community-based organizations would prefer that a facility be shut down and for investment be focused instead on cleaner production processes, such as renewable electricity.

Construction of pipelines often involves setting up work camps in remote areas. In Canada and the United States, oil and gas pipeline construction in remote communities is associated with social harms including sexual violence, and this history has led some Indigenous communities to oppose construction of CO_2 pipelines.

1.1.6 Cost

Project cost, low technology readiness levels in capture technologies, and a lack of revenue streams are among the main reasons for CCS projects to stop. A commercial-scale project typically requires an upfront capital investment of up to several billion dollars. According to the U.S. Environmental Protection Agency, CCS would increase the cost of electricity generation from coal plants by $7 to $12/MWh. The cost of CCS varies greatly by CO_2 source. If the facility produces a gas mixture with a high concentration of CO_2, as is the case for natural gas processing, it can be captured and compressed for USD 15–25/tonne. Power plants, cement plants, and iron and steel plants produce more dilute gas streams, for which the cost of capture and compression is USD 40–120/tonne CO_2. In the United States, the cost of onshore pipeline transport is in the range of USD 2–14/tonne CO_2, and more than half of onshore storage capacity is estimated to be available below USD 10/tonne CO_2. CCS implementations involve multiple technologies that are highly customized to each site, which limits the industry's ability to reduce costs through learning-by-doing.

1.1.7 Role in Climate Change Mitigation

Comparison with other Mitigation Options

Compared to other options for reducing emissions, CCS is very expensive. For instance, removing CO_2 in fossil fuel power plants increases costs by USD $50–$200 per tonne of CO_2 removed. There are many ways to reduce emissions that cost less than USD $20 per tonne of avoided CO_2 emissions. Options that have far more potential to reduce emissions at lower cost than CCS include public transit, electric vehicles, and various energy efficiency measures. Wind and solar power are often the lowest-cost ways to produce electricity, even when compared to power plants that do not use CCS. The dramatic fall in the costs of renewable power and batteries has made it difficult for fossil fuel plants with CCS to be cost-competitive.

Priority Uses

In the literature on climate change mitigation, CCS is described as having a small but critical role in reducing greenhouse gas emissions. The IPCC estimated in 2014 that forgoing CCS altogether would make it 138% more expensive to keep global warming within 2 degrees Celsius. Excessive reliance on CCS as a mitigation tool would also be costly and technically unfeasible. According to the IEA, attempting to abate oil and gas consumption only through CCS and direct air capture would cost USD 3.5 trillion per year, which is about the same as the annual revenue of the entire oil and gas industry. Emissions are relatively difficult or expensive to abate without CCS in the following niches:

Fig: The Heidelberg Brevik Carbon Capture Facility, Scheduled to Start Operating in 2025, will be the First Commercial-Scale Use of CCS in Cement Production

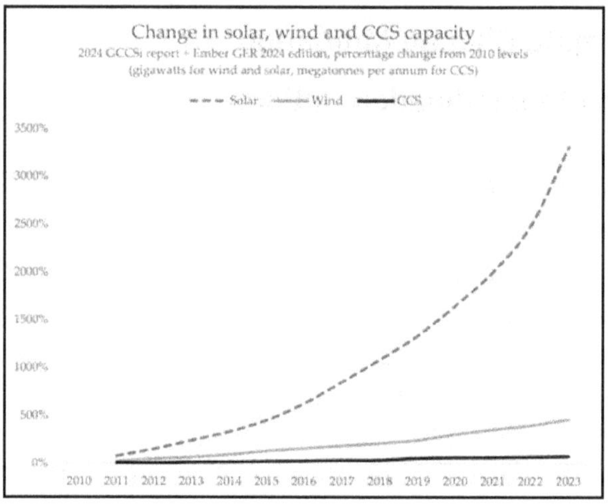

Fig: Compared to Solar and Wind Power, CCS has seen Relatively Flat Growth in Installed Capacity since 2010

- Heavy Industry: CCS is one of the few available technologies that can significantly reduce emissions associated with the production of cement, chemicals, and steel. A portion of the CO_2 emissions from these processes come from chemical reactions, in addition to emissions from burning fuels for heat. For example, approximately one third of emissions from cement making arise from burning fuels and two thirds arise from the chemical process. The Global Cement and Concrete Association say that CCS could reduce carbon emissions by 36%. Cleaner industrial processes are at varying stages of development and some have been commercialized, but are far from being widely-deployed.

- Retrofits: CCS can be retrofitted to existing coal and natural gas power plants and industrial facilities to enable the continued operation of existing plants while reducing their emissions.
- Natural Gas Processing: CCUS is the only solution to reduce the CO_2 emissions from natural gas processing. This does not reduce the emissions released when the gas is burned.
- Hydrogen: Nearly all hydrogen today is produced from natural gas or coal. Facilities can incorporate CCS to capture the CO_2 released in these processes.
- Complement to Renewable Electricity: In the IEA's scenario for net zero emissions, 251 GW of electricity worldwide are produced by coal and gas plants equipped with CCS by 2050, while 54,679 GW of electricity are produced by solar PV and wind. Although solar and wind energy are typically cheaper, power plants that burn natural gas, biomass, or coal have the advantage of being able to produce electricity in any season and any time of day, and can be dispatched at times of high demand. A small amount of power plant capacity can help to meet the growing need for system flexibility as the share of wind and solar increases. The potential for a robust power grid using 100% renewable energy has been modelled as a feasible option for many regions, which would make fossil CCS in the electricity sector unnecessary. However, this approach may be more expensive.
- Bioenergy with Carbon Capture and Storage: Bioenergy with carbon capture and storage (BECCS) is the process of extracting bioenergy from biomass and capturing and storing the CO_2 that is produced. Under some conditions, BECCS can remove carbon dioxide from the atmosphere.

The IPCC stated in 2022 that "implementation of CCS currently faces technological, economic, institutional, ecological-environmental and socio-cultural barriers." Since CCS can only be used with large, stationary emission sources, it cannot reduce the emissions from burning fossil fuels in vehicles and homes.

The IEA describes "excessive expectations and reliance" on CCS and direct air capture as a common misconception. To reach targets set in the Paris Agreement,

CCS must be accompanied by a steep decline in the production and use of fossil fuels.

Effectiveness in Reducing Greenhouse Gas Emissions

Fig: Coal Plants with CCS Usually Burn more Coal to Provide the Energy Needed for CCS Processes. This Increases the Environmental Effects of Coal Mining

When CCS is used for electricity generation, most studies assume that 85-90% of the CO_2 in the exhaust stream is captured. However, industry representatives say actual capture rates are closer to 75%, and have lobbied for government programs to accept this lower target. The potential for a CCS project to reduce emissions depends on several factors in addition to the capture rate. These factors include the amount of additional energy needed to power CCS processes, the source of the additional energy used, and post-capture leakage. The energy needed for CCS usually comes from fossil fuels whose mining, processing, and transport produce emissions. Some studies indicate that under certain circumstances the overall emissions reduction from CCS can be very low, or that adding CCS can even increase emissions relative to no capture. For instance, one study found that in the Petra Nova CCS retrofit of a coal power plant, the actual rate of emissions reduction was so low that it would average only 10.8% over a 20-year time frame.

Some CCS implementations have not sequestered carbon at their designed capacity, either for business or technical reasons. For instance, in the Shute Creek Gas Processing Facility, around half of the CO_2 that has been captured has been sold for EOR, and the other half vented to the atmosphere because it could not be profitably sold. In one year of operation of the Gorgon gas project in Australia, issues with subsurface water prevented two-thirds of captured CO_2 from being injected. A 2022 analysis of 13 major CCS projects found that most had either sequestered far less CO_2 than originally expected, or had failed entirely.

Emissions with Enhanced Oil Recovery

There is controversy over whether carbon capture followed by enhanced oil recovery is beneficial for the climate. The EOR process is energy-intensive because of the need to separate and re-inject CO_2 multiple times to minimize losses. If CO_2 losses are kept at 1%, the energy required for EOR operations results in around 0.23 tonnes of CO_2 emissions per tonne of CO_2 sequestered.

Furthermore, when the oil that is extracted using EOR is subsequently burned, CO_2 is released. If these emissions are included in calculations, carbon capture with EOR is usually found to increase overall emissions compared to not using carbon capture at all. If the emissions from burning extracted oil are excluded from calculations, carbon capture with EOR is found to decrease emissions. In arguments for excluding these emissions, it is assumed that oil produced by EOR displaces conventionally-produced oil instead of adding to the global consumption of oil.

A 2020 review found that scientific papers were roughly evenly split on the question of whether carbon capture with EOR increased or decreased emissions. The International Energy Agency's model of oil supply and demand indicates that 80% of oil produced in EOR will displace other oil on the market. Using this model, it estimated that for each tonne of CO_2 sequestered, burning the oil produced by conventional EOR leads to 0.13 tonnes of CO_2 emissions (in addition to the 0.24 tonnes of CO_2 emitted during the EOR process itself).

Pace of Implementation

As of 2023 CCS captures around 0.1% of global emissions — around 45 million tonnes of CO_2. Climate models from the IPCC and the IEA show it capturing around 1 billion tonnes of CO_2 by 2030 and several billions of tons by 2050. Technologies for CCS in high-priority niches, such as cement production, are still immature. The IEA notes "a disconnect between the level of maturity of individual CO_2 capture technologies and the areas in which they are most needed." CCS implementations involve long approval and construction times and the overall pace of implementation has historically been slow. As a result of the lack of progress, authors of climate change mitigation strategies have repeatedly reduced the role of CCS. Some observers such as the IEA call for increased commitment to CCS in order to meet targets. Other observers see the slow pace of implementation as an indication that the concept of CCS is fundamentally unlikely to succeed, and call for efforts to be redirected to other mitigation tools such as renewable energy.

1.1.8 Political Debate

CCS has been discussed by political actors at least since the start of the UNFCCC negotiations in the beginning of the 1990s, and remains a very divisive issue. Fossil fuel companies have heavily promoted CCS, framing it as an area of innovation and cost-effectiveness. Public statements from fossil fuel companies and fossil-based electric utilities ask for "recognition" that fossil fuel usage will increase in the future and suggest that CCS will allow the fossil fuel era to be extended. Their statements typically position CCS as a necessary way to tackle climate change, while not mentioning options for reducing fossil fuel use. According to the International Energy Agency, as of 2023, annual investments in the oil and gas sector are double the amount needed to produce the amount of fuel that would be compatible with limiting global warming to 1.5°C.

Fossil fuel industry representatives have had a strong presence at UN climate conferences. In these conferences, they have advocated for agreements to use language about reducing the emissions from fossil fuel use (through CCS), instead of language about reducing the use of fossil fuels. In the 2023 United Nations Climate Change Conference, at least 475 lobbyists for CCS were granted access. Many environmental NGOs such as Friends of the Earth hold strongly negative views on CCS. In surveys, environmental NGOs' importance ratings for fossil energy with CCS have been around as low as their ratings for nuclear energy. Critics see CCS as an unproven, expensive technology that will perpetuate dependence on fossil fuels. They believe other ways to reduce emissions are more effective and that CCS is a distraction. They would rather see government funds go to initiatives that are not connected to the fossil fuel industry.

Fossil Fuel Abatement

In international climate negotiations, a controversial issue has been whether to phase out use of fossil fuels generally or to phase out use of "unabated" fossil fuels. In the 2023 United Nations Climate Change Conference, an agreement was reached to phase down unabated coal use. The term abated is generally understood to mean the use of CCS, however the agreement left the term undefined. Since the terms abated and unabated were not defined, the agreement was criticized for being open to abuse.

Without a clear definition, is possible for fossil fuel use to be called "abated" if it uses CCS only in a minimal fashion, such as capturing only 30% of the emissions from a plant. The IPCC considers fossil fuels to be unabated if they are "produced and used without interventions that substantially reduce the amount of GHG emitted throughout the life-cycle; for example, capturing 90% or more from power plants, or 50–80% of fugitive methane emissions from energy supply." The intention of the IPCC definition is to require both effective CCS and deep reduction of fugitive gas emissions in order for fossil fuel emissions to qualify as being "abated."

1.1.9 Social Acceptance

The public has generally low awareness of CCS. Public support among those who are aware of CCS has tended to be low, especially compared to public support for other emission-reduction options. A frequent concern for the public is transparency, e.g. around issues such as safety, costs, and impacts. Another factor in acceptance is whether uncertainties are acknowledged, including uncertainties around potentially negative impacts on the natural environment and public health. Research indicates that engaging comprehensively with communities increases the likelihood of project success compared to projects that do not engage the public. Some studies indicate that community collaboration can contribute to the avoidance of harm within communities impacted by the project.

1.1.10 Government Programs

Almost all CCS projects operating today have benefited from government financial support, largely in the form of capital grants and – to a lesser extent – operational subsidies. Tax credits are offered in some countries. Grant funding has played a particularly important role in projects coming online since 2010, with 8 out of 15 projects receiving grants ranging from around USD 55 million (AUD 60 million) in the case of Gorgon in Australia to USD 840 million (CAD 865 million) for Quest in Canada. An explicit carbon price has supported CCS investment in only two cases to date: the Sleipner and Snøhvit projects in Norway.

North America

As a means to help boost domestic oil production, the US federal tax code has had some sort of incentive for enhanced oil recovery since 1979, when crude oil was still under federal price controls. A 15 percent tax credit was codified with the U.S. Federal EOR Tax Incentive in 1986, and oil production from EOR using CO_2 subsequently grew rapidly. In the U.S., the 2021 Infrastructure Investment and Jobs Act designates over $3 billion for a variety of CCS demonstration projects. A similar amount is provided for regional CCS hubs that focus on the broader capture, transport, and either storage or use of captured CO_2. Hundreds of millions more are dedicated annually to loan guarantees supporting CO_2 transport infrastructure.

The Inflation Reduction Act of 2022 (IRA) updates tax credit law to encourage the use of carbon capture and storage. Tax incentives under the law provide up to $85/tonne for CO_2 capture and storage in saline geologic formations or up to $60/ tonne for CO_2 used for enhanced oil recovery. The Internal Revenue Service relies on documentation from the corporation to substantiate claims on how much CO_2 is being sequestered, and does not perform independent investigations.

In 2020, a federal investigation found that claimants for the 45Q tax credit failed to document successful geological storage for nearly $900 million of the $1 billion they had claimed. In 2023 the US EPA issued a rule proposing that CCS be required in order to achieve a 90% emission reduction for existing coal-fired and natural gas power plants. That rule would become effective in the 2035-2040 time period. For natural gas power plants, the rule would require 90 percent capture of CO_2 using CCS by 2035, or co-firing of 30% low-GHG hydrogen beginning in 2032 and co-firing 96% low-GHG hydrogen beginning in 2038. Within the US, although the federal government may fully or partially fund CCS pilot projects, local or community jurisdictions would likely administer CCS project siting and construction. CO_2 pipeline safety is overseen by the Pipeline and Hazardous Materials Safety Administration, which has been criticized as being underfunded and understaffed.

Canada established a tax credit for CCS equipment for 2022–2028. The credit is 50% for CCS capture equipment and 37.5% for transportation and storage equipment. The Canadian Association of Petroleum Producers had asked for a 75% credit. The federal tax credit was expected to cost the government CAD $2.6 billion over 5 years; in 2024 the Parliamentary Budget Officer estimated it would cost CAD $5.7 billion. Saskatchewan extended its 20 per cent tax credit under the province's Oil Infrastructure Investment Program to pipelines carrying CO_2.

Europe

In Norway, CCS has been part of a strategy to make fossil fuel exports compatible with national emission-reduction goals. In 1991, the government introduced a tax on CO_2 emissions from offshore oil and gas production. This tax, combined with favorable and well-understood site geology, was a reason Equinor chose to implement CCS in the Sleipner and Snøhvit gas fields. In 2022, Denmark announced up to €5 billion in subsidies for CCS, aiming to reduce emissions by 0.9Mt of CO_2 by 2030. In the UK the CCUS roadmap outlines joint government and industry commitments to the deployment of CCUS and sets out an approach to delivering four CCUS low carbon industrial clusters, capturing 20–30 $MtCO_2$ per year by 2030. In September 2024 the UK government announced £21.7bn of subsidy over 25 years for the HyNet CCS and blue hydrogen scheme in Merseyside and the East Coast Cluster scheme in Teesside.

1.2 THE ROLE OF CCS IN CLIMATE CHANGE MITIGATION

Climate change is one of the most urgent and complex challenges facing the modern world. It is driven by the accumulation of greenhouse gases (GHGs) in the atmosphere, particularly carbon dioxide (CO_2), which is released through the burning of fossil fuels, industrial processes, and deforestation. As countries strive to reduce their carbon footprints and transition toward more sustainable energy systems, Carbon Capture and Storage (CCS) has emerged as a promising technological solution.

CCS involves capturing CO_2 emissions from sources such as power plants and industrial facilities, transporting the captured CO_2 to a storage site, and injecting it deep underground for long-term isolation from the atmosphere. Carbon Capture and Storage is a critical technology in the global effort to mitigate climate change. It offers a practical and effective way to reduce emissions from fossil fuel use and industrial processes, and it has the potential to deliver negative emissions when combined with bioenergy or direct air capture. While CCS is not without its challenges, including high costs and public acceptance issues, it remains an essential component of a comprehensive climate strategy. With the right policies, investments, and technological innovations, CCS can play a significant role in helping the world achieve its climate targets and transition toward a sustainable and low-carbon future.

1.2.1 CCS Process

The CCS process can be divided into three main stages: capture, transport, and storage. During the capture phase, CO_2 is separated from gases produced in electricity generation or industrial processes. There are three primary capture methods: pre-combustion, post-combustion, and oxy-fuel combustion. Post-combustion capture is the most commonly used and involves extracting CO_2 from flue gases after fossil fuels are burned. In the transport phase, the captured CO_2 is usually compressed and transported via pipelines, although it can also be transported by ship, especially over long distances. The final stage is storage, where CO_2 is injected into deep geological formations such as depleted oil and gas fields or deep saline aquifers. These formations are chosen for their ability to securely contain CO_2 for thousands of years, minimizing the risk of leakage.

1.2.2 Importance of CCS in Global Climate Goals

CCS plays a crucial role in achieving global climate goals, especially those outlined in the Paris Agreement, which aims to limit global temperature rise to well below 2 degrees Celsius above pre-industrial levels. According to the Intergovernmental Panel on Climate Change (IPCC), limiting warming to 1.5 degrees Celsius would be nearly impossible without CCS. This is because certain industrial processes, such as cement and steel production, are difficult to decarbonize through renewable energy alone. CCS provides a viable pathway to reduce emissions from these sectors. Moreover, when combined with bioenergy (in a process called BECCS – Bioenergy with Carbon Capture and Storage), CCS can result in negative emissions, effectively removing CO_2 from the atmosphere.

1.2.3 Current Applications and Projects Worldwide

CCS is already being implemented in various parts of the world, although on a relatively small scale. There are currently around 30 commercial CCS facilities either operating or under development, primarily in countries like the United States, Canada, Norway, and Australia. Notable projects include the Boundary Dam project in Canada, which retrofitted a coal-fired power plant with CCS technology, and the Sleipner project in Norway, which has been storing CO_2 in a saline aquifer beneath the North Sea since 1996. These projects have demonstrated that CCS can be technically feasible and effective, though challenges remain in scaling up the technology to the level required for meaningful climate impact.

1.2.4 Advantages of CCS Technology

One of the main advantages of CCS is its potential to significantly reduce CO_2 emissions from existing infrastructure. This is particularly important for regions that rely heavily on fossil fuels and may face economic or political obstacles in transitioning to renewables in the short term. CCS allows these regions to reduce emissions while continuing to utilize their current energy systems. Additionally, CCS can support a just transition by preserving jobs in traditional energy sectors and creating new employment opportunities in the CCS value chain. From a technological perspective, CCS is also highly versatile and can be applied to various industries beyond power generation, including chemicals, refining, and hydrogen production.

1.2.5 Challenges and Limitations of CCS

Despite its potential, CCS faces several significant challenges. One of the primary barriers is the high cost associated with capturing, transporting, and storing CO_2. These costs can make CCS less economically attractive compared to other mitigation options, especially in the absence of strong carbon pricing or policy support. Furthermore, public perception and acceptance of CCS can be an obstacle, particularly regarding the safety and environmental risks associated with underground storage. There is also concern that an overreliance on CCS might delay the adoption of renewable energy technologies and perpetuate the use of fossil fuels. Ensuring robust monitoring and regulation of storage sites is essential to address these concerns and build public trust in CCS as a safe and effective solution.

1.2.6 Policy and Regulatory Support for CCS

The successful deployment of CCS on a large scale requires strong policy frameworks and regulatory support. Governments must create incentives for investment in CCS through measures such as carbon pricing, tax credits, and funding for research and development. For example, the United States has introduced the 45Q tax credit, which provides financial incentives for companies that capture and store CO_2. In Europe, the EU Innovation Fund supports large-scale CCS projects as part of its broader climate strategy. Effective regulation is also needed to ensure that CO_2 is stored safely and permanently. This includes rigorous site selection, monitoring, and verification procedures, as well as legal frameworks that address liability and ownership of stored CO_2.

1.2.7 Role of CCS in a Diverse Energy Future

CCS should not be viewed as a silver bullet for climate change mitigation but rather as part of a broader portfolio of solutions. It complements other strategies such as energy efficiency, renewable energy deployment, and behavioral change. In particular, CCS is well-suited to play a key role in decarbonizing the so-called "hard-to-abate" sectors that cannot easily switch to electricity or hydrogen. Furthermore, the development of Direct Air Capture (DAC) technologies, which remove CO_2 directly from the atmosphere, represents an exciting extension of CCS. While still in the early stages, DAC combined with geological storage could help achieve net-zero or even net-negative emissions in the future.

1.2.8 Future of CCS: Innovation and Collaboration

Innovation will be crucial to reducing the cost and improving the efficiency of CCS technologies. Advances in materials science, capture techniques, and digital monitoring tools can make CCS more economically viable and scalable. Collaboration between governments, industry, and research institutions is also essential to accelerate deployment. International cooperation can facilitate the sharing of best practices, harmonization of standards, and development of cross-border CO_2 transport and storage infrastructure. As the urgency to reduce emissions grows, there is increasing recognition of the need to invest in all available mitigation tools, including CCS.

1.3 HISTORICAL DEVELOPMENT OF CCS TECHNOLOGIES

Carbon Capture and Storage (CCS) technologies have evolved over several decades, shaped by advancements in science, engineering, environmental policy, and the urgent global need to address climate change. The journey of CCS is deeply connected with the industrial revolution, fossil fuel usage, and the eventual realization that carbon dioxide emissions significantly contribute to global warming. The historical development of CCS technologies is marked by key scientific discoveries, pilot projects, commercial implementation, and international collaborations aimed at reducing greenhouse gas emissions.

The historical development of Carbon Capture and Storage technologies reflects a complex interplay of scientific discovery, industrial innovation, environmental awareness, and policy action. From its roots in gas separation and enhanced oil recovery to its current role in climate change mitigation, CCS has evolved into a multifaceted solution capable of addressing emissions from a wide range of sectors. While challenges remain, especially in terms of cost and scalability, the progress made over the past decades demonstrates the potential of CCS to contribute significantly to global climate goals. With continued investment and innovation, CCS can become a cornerstone of a low-carbon future, offering a bridge between current energy systems and the sustainable transitions required to safeguard the planet.

1.3.1 Scientific Origins and Early Industrial Practices

The basic science behind carbon dioxide capture and storage dates back to the early 20th century when researchers first explored methods to separate and use gases from industrial processes. Early industrial applications involved capturing CO_2 for use in the food and beverage industry and in enhanced oil recovery (EOR). By the 1930s, CO_2 was being injected into oil fields to increase crude oil extraction, although this was not intended for climate purposes.

These initial uses laid the foundation for later CCS technologies, particularly in understanding how gases could be injected into underground formations and managed over time. During the mid-20th century, chemical engineers developed more refined methods for gas separation, particularly using amine-based solvents to remove CO_2 from natural gas. These techniques, although not designed for climate mitigation, became essential components of future carbon capture processes.

1.3.2 Environmental Movement and the Birth of Modern CCS

The environmental movement of the 1960s and 1970s brought attention to the harmful effects of air pollution and fossil fuel use. Scientists began to study the atmospheric concentration of greenhouse gases, with Charles David Keeling's measurements showing a steady increase in CO_2 levels. These findings, along with emerging climate models, helped raise awareness about the role of human activities in global warming.

By the 1980s, the concept of capturing and storing CO_2 as a means to mitigate climate change gained attention in academic and policy circles. Researchers began to consider the possibility of adapting existing gas separation and injection technologies for environmental use. It was during this period that the term "carbon capture and storage" started to appear in scientific literature, signaling the beginning of modern CCS thinking.

1.3.3 Pilot Projects and Early Demonstrations

The 1990s marked a turning point in the practical development of CCS, with the launch of the world's first large-scale CCS project at the Sleipner gas field in Norway in 1996. Operated by Statoil (now Equinor), the project captured CO_2 from natural gas production and injected it into a saline aquifer beneath the North Sea. This project demonstrated that geological storage was both technically feasible and environmentally secure. It also provided a valuable learning platform for monitoring techniques, regulatory frameworks, and public communication strategies. The success of the Sleipner project inspired similar pilot projects across the globe, including initiatives in Canada, the United States, and Australia. These early projects focused on improving capture efficiency, testing storage integrity, and developing cost-effective solutions for various industries.

1.3.4 Commercial Deployment and Policy Support

As concerns over climate change intensified in the early 2000s, governments and industries began to invest more heavily in CCS research and infrastructure. The United States Department of Energy, the European Commission, and other national agencies provided funding for large-scale demonstration projects aimed at testing CCS under real-world conditions. The Boundary Dam project in Saskatchewan, Canada, launched in 2014, became the first commercial coal-fired power plant retrofitted with CCS technology.

It captured up to 1 million tonnes of CO_2 annually and stored it underground or used it for EOR. Around the same time, the Gorgon project in Australia began injecting CO_2 from a liquefied natural gas (LNG) facility into deep geological formations. These projects highlighted both the potential and the challenges of CCS, particularly regarding cost, technical complexity, and regulatory requirements. In parallel, global climate agreements, such as the Kyoto Protocol and later the Paris Agreement, emphasized the importance of reducing greenhouse gas emissions and supported the inclusion of CCS in national climate strategies.

1.3.5 Technological Innovations and Industry Diversification

Over the past two decades, there have been significant technological advancements in all three components of CCS: capture, transport, and storage. In capture technology, improvements have been made in solvent chemistry, membrane separation, and cryogenic techniques, increasing efficiency and reducing energy penalties. Transport infrastructure, particularly pipeline networks, has expanded in key regions such as North America and Europe, facilitating the movement of CO_2 from capture sites to storage locations. In terms of storage, better geological modeling, real-time monitoring tools, and risk assessment protocols have enhanced the safety and reliability of CO_2 injection. Additionally, CCS has diversified beyond power generation to include applications in cement, steel, hydrogen production, and waste-to-energy facilities. This diversification reflects a broader understanding of the need to decarbonize industrial sectors that cannot easily transition to renewable energy.

1.3.6 Integration with Negative Emissions and Emerging Technologies

The concept of negative emissions has further expanded the role of CCS in climate mitigation. Bioenergy with Carbon Capture and Storage (BECCS) combines biomass energy production with CCS to achieve net-negative emissions. This means that

CO_2 absorbed by plants during growth is captured and stored rather than released back into the atmosphere, effectively removing CO_2 from the air. Similarly, Direct Air Capture (DAC) technologies, which extract CO_2 directly from ambient air, have begun to integrate with CCS storage techniques. Though still in the early stages, these technologies offer promising avenues for removing legacy emissions and achieving net-zero targets. Several startups and research institutions are actively developing scalable DAC solutions, with pilot facilities in operation in Europe and North America.

1.3.7 Global Cooperation and Future Outlook

The development of CCS has increasingly become a global effort, with international collaborations playing a key role in sharing knowledge, setting standards, and financing projects. Organizations such as the Global CCS Institute, the International Energy Agency (IEA), and the Carbon Sequestration Leadership Forum (CSLF) have facilitated cooperation among governments, industries, and research bodies. These efforts have helped build public trust, streamline permitting processes, and foster innovation. Looking ahead, the future of CCS depends on continued policy support, financial incentives, and public-private partnerships. Carbon pricing mechanisms, such as carbon taxes and cap-and-trade systems, are critical to making CCS economically viable. Governments must also invest in infrastructure development, such as CO_2 pipelines and storage hubs, to reduce logistical barriers and enable large-scale deployment.

1.4 THE NEED FOR CCS IN ACHIEVING GLOBAL CLIMATE TARGETS

1.4.1 Paris Agreement and Net-Zero Emissions

The Paris Agreement stands as a landmark in the global response to climate change, marking a unified commitment by nations to tackle the escalating environmental crisis. Adopted in December 2015 at the 21st Conference of the Parties (COP21) of the United Nations Framework Convention on Climate Change (UNFCCC), the Paris Agreement brought together 196 countries in a legally binding international treaty. Its primary objective is to limit global warming to well below 2 degrees Celsius above pre-industrial levels, with efforts to restrict it further to 1.5 degrees. One of the key mechanisms to achieve this temperature goal is the collective pursuit of net-zero emissions, a state where the amount of greenhouse gases emitted is balanced by those removed from the atmosphere.

The Paris Agreement and the concept of net-zero emissions represent a transformative shift in the global approach to climate change. By establishing a universal framework for climate action and emphasizing long-term goals, the agreement has catalyzed ambition and fostered international solidarity. Achieving net-zero emissions is essential to limiting global warming and avoiding the most severe impacts of climate change. While the journey to net-zero is fraught with challenges, it is also filled with opportunities for innovation, resilience, and sustainable development. Success will require unwavering political commitment, substantial investment, inclusive policies, and global cooperation. As the world navigates this critical decade of climate action, the pursuit of net-zero stands as a guiding principle for securing a safe and livable future for generations to come.

Foundation and Objectives of the Paris Agreement

The Paris Agreement emerged from years of international negotiation and scientific consensus on the urgent need to address climate change. Unlike its predecessor, the Kyoto Protocol, which imposed binding targets only on developed nations, the Paris Agreement adopts a more inclusive and flexible framework, encouraging all countries, regardless of their level of development, to contribute to climate action. Its key objective is to strengthen the global response to the threat of climate change by keeping the rise in global average temperature to well below 2 degrees Celsius. It further aims to pursue efforts to limit the temperature increase to 1.5 degrees, a threshold that scientists associate with significantly lower risks of catastrophic climate impacts. To achieve these goals, the Paris Agreement requires countries to submit Nationally Determined Contributions (NDCs), outlining their plans to reduce greenhouse gas emissions and adapt to climate change. These contributions are to be updated every five years, with each successive NDC expected to reflect greater ambition. The agreement also calls for enhanced transparency, climate finance to assist developing countries, and the strengthening of adaptive capacities.

Net-Zero Emissions

Net-zero emissions, also known as carbon neutrality, refers to a state where the total greenhouse gas emissions produced by human activities are balanced by an equivalent amount of emissions removed from the atmosphere. Achieving net-zero does not imply eliminating all emissions but rather compensating for any remaining emissions through removal strategies. These may include natural carbon sinks like forests and oceans or technological solutions such as Carbon Capture and Storage (CCS) and Direct Air Capture (DAC). The net-zero concept has gained widespread recognition as an essential pathway to achieving the temperature goals of the Paris Agreement. The Intergovernmental Panel on Climate Change (IPCC) has emphasized that reaching net-zero CO_2 emissions by around 2050 is crucial to limit warming to 1.5 degrees Celsius. Delays in action would necessitate even more aggressive reductions and greater reliance on negative emissions technologies later, which could pose economic and technical difficulties.

Commitments by Countries towards Net-Zero

Since the adoption of the Paris Agreement, many countries have declared their intention to achieve net-zero emissions, setting specific target years for reaching this goal. The European Union, for instance, has committed to becoming climate-neutral by 2050, enshrining this target in law through the European Climate Law. Similarly, the United Kingdom was among the first major economies to legislate a net-zero target for 2050. China, the world's largest emitter, has pledged to achieve carbon neutrality by 2060, while the United States rejoined the Paris Agreement under the Biden administration and announced a goal to reach net-zero by 2050. These commitments are important milestones in global climate action, demonstrating growing political will and international solidarity. However, achieving net-zero requires translating pledges into robust policies, regulations, and investments. Many countries still face significant gaps between their stated ambitions and the actions needed to fulfill them, underscoring the importance of continuous progress and accountability.

Strategies for Achieving Net-Zero

Reaching net-zero emissions involves a multifaceted approach that spans across sectors and scales. The transformation of the energy sector is central to this goal, as it remains the largest source of global emissions. This transformation entails a rapid shift from fossil fuels to renewable energy sources such as wind, solar, and hydroelectric power. Improving energy efficiency in buildings, appliances, and transportation can also significantly reduce emissions. Electrification of transport and industry, supported by decarbonized electricity, forms another pillar of net-zero strategies. In sectors where emissions are harder to eliminate, such as aviation, shipping, and heavy industry, low-carbon fuels like hydrogen and sustainable biofuels offer viable solutions. Additionally, nature-based solutions, including afforestation, reforestation, and improved land management, play a vital role in sequestering carbon. Technological innovation is also essential. Carbon removal technologies such as BECCS and DAC are expected to contribute to offsetting residual emissions, especially in the later stages of the transition. Governments and industries must invest in research, development, and deployment to make these technologies scalable and cost-effective.

Challenges in the Path to Net-Zero

Despite growing momentum, achieving net-zero emissions presents numerous challenges. One major hurdle is the political and economic cost of rapid decarbonization. Phasing out fossil fuels, reforming subsidies, and transforming industrial practices can encounter resistance from entrenched interests and affected communities. Policymakers must balance environmental goals with economic stability and social equity. Another challenge lies in the availability and scalability of carbon removal technologies. While some methods have shown promise, they remain expensive or technically unproven at large scale. Overreliance on future technological breakthroughs could delay immediate action and jeopardize climate targets. Equity and climate justice are also critical considerations. Developing countries, which have contributed the least to historical emissions, are often the most vulnerable to climate impacts. These nations require financial and technical support to pursue low-carbon development pathways while addressing poverty, infrastructure deficits, and capacity constraints. Ensuring a fair and inclusive transition is essential for the legitimacy and effectiveness of global climate efforts.

Global Cooperation and Monitoring Progress

Global cooperation is fundamental to the success of the Paris Agreement and the net-zero transition. The agreement's framework of NDCs, regular stocktakes, and transparency mechanisms is designed to foster accountability and encourage collective ambition. The Global Stocktake, conducted every five years, assesses progress towards the goals of the agreement and informs future NDCs. Climate finance is another cornerstone of international cooperation. Developed countries have committed to mobilizing $100 billion annually to support climate action in developing nations. While progress has been made, many countries argue that the pledged funds remain insufficient or delayed. Meeting and surpassing this financial commitment is essential to building trust and enabling equitable climate action. The role of non-state actors, including cities, regions, businesses, and civil society, is also crucial. Many corporations have set net-zero targets, and subnational governments are implementing climate plans aligned with national and global goals. These actors can drive innovation, build public support, and enhance policy implementation at the local level.

1.4.2 Role of CCS in Deep Decarbonization

As the world strives to tackle the urgent threat of climate change, deep decarbonization has become a core strategy for achieving net-zero greenhouse gas emissions by mid-century. Deep decarbonization refers to the profound and economy-wide reduction of carbon emissions, typically by 80% to 100% from current levels. While the deployment of renewable energy sources, energy efficiency improvements, and behavioral changes are essential components of this strategy, they are not sufficient on their own. Carbon Capture and Storage (CCS) plays a pivotal role in ensuring that deep decarbonization is feasible, especially in hard-to-abate sectors such as heavy industry, energy-intensive manufacturing, and fossil-based power generation. Carbon Capture and Storage is an essential enabler of deep decarbonization. By addressing emissions from hard-to-abate sectors, supporting low-carbon hydrogen production, enabling negative emissions, and complementing renewable energy, CCS forms a critical pillar in the global effort to combat climate change. While challenges remain, technological advancements, supportive policies, and growing international commitment are paving the way for CCS to become a key part of a sustainable, low-carbon future. As the world moves closer to net-zero targets, the role of CCS will only grow in importance, offering a realistic and effective path toward a decarbonized global economy.

Deep Decarbonization and the Need for CCS

Deep decarbonization is an ambitious goal that requires transforming the global energy and industrial systems in order to reduce carbon dioxide and other greenhouse gas emissions to net-zero. This transformation demands a diverse portfolio of solutions, including electrification of transportation and industry, large-scale deployment of renewable energy, development of energy storage technologies, and behavioral changes in consumption patterns. However, many emissions sources are either technically challenging or prohibitively expensive to eliminate using conventional renewable energy solutions. In such cases, CCS becomes indispensable. Carbon Capture and Storage is a suite of technologies designed to capture carbon dioxide from large point sources such as power plants and industrial facilities, transport it via pipelines or other means, and store it safely underground in geological formations. By preventing CO_2 from entering the atmosphere, CCS directly reduces emissions from sectors that are otherwise difficult to decarbonize. It also enables the production of low-carbon hydrogen and facilitates negative emissions through processes like Bioenergy with Carbon Capture and Storage (BECCS).

CCS in Hard-to-Abate Sectors

Heavy industries such as cement, steel, aluminum, and chemicals are critical to modern economies, yet they contribute significantly to global CO_2 emissions. These sectors rely on high-temperature heat and specific chemical reactions that inherently produce carbon emissions. For example, the process of producing clinker, a key ingredient in cement, involves the calcination of limestone, which releases large quantities of CO_2. In such cases, CCS is the only technology currently available that can capture and reduce emissions at scale without altering the fundamental industrial processes.

Similarly, the steel industry, which predominantly uses blast furnaces fueled by coal, faces substantial decarbonization challenges. While alternative methods such as hydrogen-based direct reduced iron (DRI) are being developed, they are still in early stages of commercialization. CCS provides an immediate and effective solution by capturing emissions from existing facilities, allowing the industry to transition gradually without compromising production. The chemical industry, including the production of ammonia, fertilizers, and petrochemicals, also generates large emissions due to the use of fossil feedstocks. Integrating CCS into these operations helps reduce emissions without significant disruptions. As a result, CCS serves as a vital bridge to a low-carbon future while new processes and technologies mature.

Power Sector Decarbonization and Energy Security

In the power sector, CCS can enable deep decarbonization by reducing emissions from fossil-fuel-based generation. While renewable energy sources such as wind and solar are growing rapidly, they are intermittent by nature and require backup generation capacity. Fossil fuel power plants equipped with CCS provide a low-carbon solution to ensure grid reliability and stability, especially during periods of low renewable output. Moreover, CCS allows the continued use of existing power infrastructure, avoiding the cost and disruption associated with replacing entire fleets of plants. It also plays a role in repurposing retired coal plants by retrofitting them with carbon capture systems, thereby extending their lifespan in a climate-compatible way. This contributes to energy security, particularly in countries where fossil fuels are still a major part of the energy mix.

Enabling Low-Carbon Hydrogen Production

Hydrogen is widely regarded as a key component of deep decarbonization strategies due to its versatility and ability to decarbonize sectors such as transportation, industry, and heating. However, most hydrogen today is produced from natural gas through a process known as steam methane reforming (SMR), which emits large amounts of CO_2. By integrating CCS with SMR, it is possible to produce what is known as "blue hydrogen," which significantly reduces the carbon footprint compared to conventional methods. Blue hydrogen serves as a transitional solution while green hydrogen, produced through electrolysis using renewable electricity, becomes more economically viable. Thus, CCS supports the development of a low-carbon hydrogen economy and helps bridge the gap between current technologies and future innovations.

Role in Achieving Negative Emissions

Another vital role of CCS in deep decarbonization is its contribution to negative emissions. As certain emissions sources may be impossible to eliminate entirely, achieving net-zero will require the active removal of CO_2 from the atmosphere. Technologies such as BECCS and Direct Air Capture with Storage (DACCS) rely on CCS to permanently store the captured carbon underground. BECCS involves using biomass to produce energy and capturing the CO_2 emitted during combustion or gasification. Since the biomass absorbed CO_2 during its growth, capturing and storing it results in a net-negative emissions process. DACCS, although still in early stages of deployment, captures CO_2 directly from the ambient air and stores it similarly. These technologies are expected to play a critical role in offsetting residual emissions and helping achieve climate targets.

Technological Advancements and Innovation

The effectiveness of CCS in supporting deep decarbonization depends on continuous technological innovation. Advances in capture technologies, such as improved solvents, membranes, and sorbents, are reducing the energy requirements and costs of CO_2 capture. New approaches, including solid oxide fuel cells and oxy-fuel combustion, offer potential efficiency gains and lower emissions. Transport and storage technologies are also evolving, with the development of dedicated CO_2 pipelines, shipping methods, and storage hubs. Deep saline aquifers, depleted oil and gas reservoirs, and basalt formations are among the geological options being explored for long-term CO_2 storage. Enhanced oil recovery (EOR) using captured CO_2 provides a near-term economic incentive for CCS deployment, although its climate benefits depend on careful regulation.

Policy Support and Financial Incentives

For CCS to contribute meaningfully to deep decarbonization, supportive policies and financing mechanisms are essential. The high capital costs of capture equipment, infrastructure, and storage sites often deter private investment. Government intervention through carbon pricing, tax credits, grants, and regulatory frameworks can bridge this gap and drive CCS deployment. Examples of successful policy mechanisms include the 45Q tax credit in the United States, which offers financial incentives for each ton of CO_2 captured and stored. In Europe, the EU Innovation Fund and carbon contracts for difference provide similar support. Public-private partnerships, international collaboration, and multilateral funding can further accelerate the development of CCS projects, particularly in developing countries.

Challenges and Public Perception

Despite its potential, CCS faces several challenges. High upfront costs, limited infrastructure, regulatory hurdles, and public skepticism about safety and environmental risks have slowed its widespread adoption. Concerns about induced seismicity, leakage, and the long-term stability of storage sites need to be addressed through robust monitoring and transparent communication. Public understanding and acceptance of CCS are crucial for its successful implementation. Building trust through education, stakeholder engagement, and clear demonstration of safety and efficacy can help overcome resistance. Integrating CCS within broader decarbonization strategies and ensuring that it complements rather than delays other solutions is also vital.

1.5 OVERVIEW OF CCS PROCESS: CAPTURE, TRANSPORT, AND STORAGE

Carbon Capture and Storage (CCS) is a critical climate mitigation technology that aims to reduce carbon dioxide (CO_2) emissions from large industrial sources and fossil-fuel-based power plants. It involves a three-step process: capturing CO_2 at the source, transporting it to a suitable location, and storing it permanently underground. CCS plays a vital role in enabling countries to meet their emission reduction targets, particularly in sectors where emissions are difficult to eliminate entirely. As part of a broader strategy for climate stabilization, CCS supports both the transition to net-zero emissions and the potential for negative emissions when used in conjunction with bioenergy or direct air capture. Carbon Capture and Storage is a vital technology in the global effort to combat climate change. By capturing emissions from large point sources, transporting CO_2 to secure locations, and storing it underground, CCS offers a practical and scalable solution for reducing atmospheric greenhouse gases. Each step of the process—capture, transport, and storage—presents unique technical and logistical challenges, but together they form a comprehensive system for deep decarbonization. As countries work to meet ambitious climate targets and transition to net-zero emissions, the development and deployment of CCS infrastructure will be essential. With continued innovation, supportive policies, and public engagement, CCS can become a cornerstone of a sustainable and climate-resilient future.

1.5.1 Carbon Capture: The First Step in the CCS Chain

The capture phase is the first and most technically intensive step in the CCS process. It involves separating carbon dioxide from the gas streams generated during combustion or industrial processes. This separation is essential because, without it, CO_2 would escape into the atmosphere and contribute to the greenhouse effect. Capture technologies vary depending on the type of emission source and the composition of the exhaust gas. The most common methods are post-combustion, pre-combustion, and oxy-fuel combustion capture. Post-combustion capture is widely used and involves removing CO_2 from flue gases after fossil fuels have been burned. This method is particularly applicable to existing power plants and industrial facilities, making it suitable for retrofitting. The process typically uses chemical solvents, such as amines, which absorb CO_2 from the gas stream. After absorption, the solvent is heated to release pure CO_2, which can then be compressed and prepared for transport.

Pre-combustion capture, on the other hand, is primarily used in gasification-based systems. In this method, fuel is converted into a mixture of hydrogen and carbon monoxide known as syngas. The carbon monoxide is then reacted with water to form CO_2 and more hydrogen. The CO_2 is separated before combustion, while the hydrogen is used as a clean energy source. This approach is efficient and produces a concentrated stream of CO_2, making capture easier and less energy-intensive. Oxy-fuel combustion capture involves burning fossil fuels in pure oxygen instead of air. This results in a flue gas that consists mostly of CO_2 and water vapor, making it easier to separate and purify the CO_2. Though this method requires additional energy to produce oxygen, it offers a relatively simple separation process and is considered suitable for new-build facilities designed for CCS integration.

1.5.2 Carbon Transport: Moving CO_2 to Storage Sites

Once captured, the CO_2 must be transported to a suitable storage site. Transportation is a crucial link in the CCS chain, as storage sites are often located far from emission sources. The most common method of transporting CO_2 is via pipelines, which are already widely used in the oil and gas industry and have a proven safety record. CO_2 pipelines are typically constructed from carbon steel and operate at high pressures to keep the gas in a dense, supercritical state, reducing its volume and making transport more efficient. Pipeline networks can be designed as dedicated point-to-point systems, linking a single emitter to a storage site, or as shared infrastructure hubs serving multiple sources. The development of CO_2 transport infrastructure is critical for scaling up CCS, as it allows smaller emitters to access storage services through shared systems, reducing costs and increasing efficiency.

In addition to pipelines, CO_2 can also be transported by ship, particularly when dealing with offshore storage sites or regions where pipeline construction is not feasible. Shipping offers flexibility and can be integrated into existing port and tanker infrastructure. For transport by ship, CO_2 is typically liquefied at low temperatures and stored in pressurized tanks. Though shipping is more expensive than pipelines for continuous transport, it can be cost-effective for moderate volumes and for initial deployment phases. Other modes of transport, such as rail and road tankers, are generally used for small-scale or pilot projects. These methods are less suitable for large-scale CCS due to their limited capacity and higher operational costs. However, they may play a role in early-stage demonstration projects or in areas with dispersed sources.

1.5.3 Carbon Storage: Permanently Sequestering CO_2 Underground

The final step in the CCS process is the permanent storage of CO_2 in deep underground geological formations. Effective storage ensures that the captured CO_2 does not return to the atmosphere and contributes to long-term climate mitigation. There are several types of geological formations suitable for CO_2 storage, including deep saline aquifers, depleted oil and gas fields, and unmineable coal seams. Deep saline aquifers are the most abundant and widely considered option for large-scale storage. These formations contain porous rock saturated with brine and are located thousands of meters below the Earth's surface. When CO_2 is injected into these formations, it is trapped through a combination of physical and chemical mechanisms, including structural trapping under impermeable cap rock, residual trapping within rock pores, solubility trapping as CO_2 dissolves in brine, and mineral trapping as it reacts with surrounding rock to form stable carbonates. Depleted oil and gas fields offer another promising storage option. These formations have already proven capable of trapping hydrocarbons for millions of years and are often well-characterized, with existing infrastructure such as wells and pipelines. Using these fields for CO_2 storage also allows for enhanced oil recovery (EOR), a process in which CO_2 is injected to extract additional oil. While EOR can generate revenue and encourage early investment in CCS, it must be carefully managed to ensure that it delivers net emission reductions. Unmineable coal seams, which cannot be economically mined, may also be used for CO_2 storage. In these formations, CO_2 is adsorbed onto the coal surface, displacing methane that can be recovered and used as an energy source. This process, known as enhanced coal bed methane recovery, combines storage with energy production. However, it is less developed than other storage options and presents specific technical challenges. To ensure the safety and effectiveness of CO_2 storage, rigorous monitoring and verification are essential. Advanced techniques, including seismic imaging, pressure monitoring, and tracer gas analysis, are used to track the movement of CO_2 and detect any potential leaks. Regulatory frameworks and site-specific risk assessments are necessary to prevent environmental impacts and maintain public trust.

1.5.4 Integration and Deployment Challenges

Successfully implementing CCS requires the integration of capture, transport, and storage systems into a cohesive value chain. Each component must be coordinated to match capacity, timing, and location, ensuring smooth operation and cost efficiency. Large-scale deployment also faces financial, regulatory, and public acceptance challenges. High capital costs, particularly for capture equipment, are a major barrier. Incentive mechanisms such as carbon pricing, tax credits, and public investment are necessary to make CCS economically viable. Regulatory approval and permitting processes can be complex and time-consuming, especially for storage sites. Governments must establish clear legal frameworks that define liability, ownership, and monitoring responsibilities. Public perception of CCS, especially concerns about safety and long-term environmental risks, must be addressed through education, transparency, and stakeholder engagement.

CHAPTER

Technologies for Carbon Capture

2

2.1 POST-COMBUSTION CAPTURE

2.1.1 How Post-Combustion Capture Works

Post-combustion capture is a process where carbon dioxide (CO_2) is removed from the flue gas (exhaust gases) produced after fossil fuel combustion in power plants or other industrial facilities. This CO_2 is then separated, compressed, and potentially stored or used.

- Flue Gas Production: Fossil fuel combustion (like coal or natural gas) produces flue gas, which is a mix of gases including CO_2, water vapor, and other byproducts like sulfur dioxide and nitrogen oxides.
- CO_2 Separation:
  - Absorption: A common method is to use a solvent, like monoethanolamine (MEA), which selectively absorbs CO_2 from the flue gas.
  - Regeneration: The solvent, now enriched with CO_2, is then heated to release the captured CO_2 as a concentrated stream.
- CO_2 Compression and Storage/Utilization: The captured CO_2 is compressed for transportation to a storage site or for potential utilization in various industries.
  - Storage sites can be depleted oil or gas reservoirs or deep saline aquifers.
  - The captured CO_2 can also be used in various industrial processes, such as enhancing oil recovery or in the manufacturing of products like plastics, methanol, or urea.
- Advantages:
  - Post-combustion capture can be retrofitted into existing power plants, making it a cost-effective option.
  - It is applicable to various industries beyond power plants, including cement and steel production.
  - It can reduce a power plant's CO_2 emissions by a significant amount (up to 80-90%).

- Challenges:
  - The process requires a large amount of equipment and space.
  - Solvent degradation and losses can be a concern.
  - The energy penalty (additional energy needed to operate the capture process) can impact the efficiency of power plants.

2.1.2 Solvent-Based Capture Systems

Solvent-based capture systems are among the most mature and widely deployed technologies for carbon dioxide (CO_2) removal in carbon capture and storage (CCS) processes. These systems are specifically designed to separate CO_2 from gas streams produced by fossil fuel combustion and industrial activities. The technology primarily involves the use of chemical solvents that react with CO_2, allowing for its extraction from mixed gases. Solvent-based methods are particularly significant in post-combustion capture scenarios, where CO_2 must be separated from the flue gas emitted after the fuel has been burned. Solvent-based capture systems represent a vital and well-established approach to carbon dioxide removal in the context of carbon capture and storage. By enabling the efficient separation of CO_2 from industrial and power sector emissions, these systems support global climate mitigation strategies and the transition to a low-carbon future. Despite challenges related to energy consumption, environmental impact, and cost, ongoing research and technological advancements continue to enhance their performance and commercial viability. As part of a broader portfolio of climate solutions, solvent-based capture systems are poised to play a leading role in the global effort to reduce greenhouse gas emissions and achieve sustainable development goals.

Principle of Operation in Solvent-Based Systems

The fundamental principle of solvent-based capture is the selective absorption of CO_2 from a gas mixture into a liquid solvent. In most applications, the process occurs in two main steps: absorption and regeneration. During absorption, the flue gas containing CO_2 is brought into contact with the solvent in an absorber column. The solvent binds with the CO_2 to form a chemical compound, typically at near atmospheric pressure and moderate temperatures. After absorption, the CO_2-rich solvent is transferred to a regeneration unit or stripper, where it is heated to release the absorbed CO_2. The stripped CO_2 is then compressed for transportation and storage, while the regenerated solvent is recycled back into the absorber for reuse. This cyclic process allows for continuous CO_2 capture and represents a scalable and flexible solution adaptable to various industrial settings.

Types of Solvents Used in CO_2 Capture

Several categories of solvents are used in CO_2 capture, each with distinct chemical properties, absorption efficiencies, and regeneration energy requirements. The most commonly used solvents are amine-based compounds, which are favored for their strong chemical reactivity with CO_2 and proven industrial track record. Among these, monoethanolamine (MEA) is the most extensively studied and used. MEA is a primary amine that reacts rapidly with CO_2 to form a carbamate compound, which is then decomposed during the regeneration process to release pure CO_2.Other amines used include diethanolamine (DEA), methyldiethanolamine (MDEA), and piperazine (PZ), each offering different trade-offs in terms of reaction rates, absorption capacities, and thermal stability.

Secondary and tertiary amines, such as MDEA, are often blended with primary amines to improve energy efficiency and lower degradation rates. Recently, researchers have developed advanced solvents, such as sterically hindered amines and aqueous ammonia, which provide higher CO_2 loading capacities and lower regeneration energy requirements. Additionally, non-aqueous solvents, ionic liquids, and biphasic solvents are being explored for their potential to overcome limitations associated with conventional solvents.

Application in Power Plants and Industrial Settings

Solvent-based capture systems are primarily applied in post-combustion CO_2 capture from coal and natural gas-fired power plants. These plants produce large volumes of flue gas containing relatively low concentrations of CO_2 mixed with nitrogen, oxygen, and water vapor. Solvent-based systems are particularly suited to this application because they can be retrofitted to existing infrastructure and offer a reliable method for extracting CO_2 from dilute gas streams. Beyond power generation, solvent-based systems are also employed in industrial facilities such as cement plants, steel mills, and refineries.

These industries are characterized by high process emissions and cannot easily switch to renewable energy sources due to the nature of their operations. Solvent-based technologies provide a practical means of reducing emissions from these sources, contributing significantly to overall decarbonization efforts. Their modular design and scalability make them adaptable to a wide range of industrial emission profiles.

Efficiency and Energy Requirements

A critical consideration in the implementation of solvent-based capture systems is their energy requirement, particularly during the regeneration phase. The process of heating the solvent to release CO_2 consumes a significant amount of thermal energy, which often comes from the same power plant or industrial source. This energy demand, known as the reboiler duty, can reduce the overall efficiency of the plant, leading to what is termed an "energy penalty." To mitigate this, extensive research is being conducted to develop solvents with lower heat capacities and better thermal properties. The use of blended amines, process optimization, and heat integration strategies are helping to reduce energy consumption. Nonetheless, the energy penalty remains a major challenge, especially for large-scale deployment, and continues to be a focus for technological improvement.

Environmental and Operational Considerations

While solvent-based systems are effective in capturing CO_2, they also present certain environmental and operational challenges. One concern is solvent degradation, which occurs due to exposure to oxygen, impurities in the flue gas, and high operating temperatures. Degradation not only reduces solvent performance but also leads to the formation of corrosive and potentially toxic by-products that must be managed carefully. Another issue is solvent loss through evaporation and entrainment in the treated gas. This can result in emissions of volatile organic compounds (VOCs) and necessitates the use of gas cleaning equipment, such as water wash sections or demisters, to minimize environmental impact. Additionally, the corrosive nature of some solvents requires careful material selection and maintenance practices to ensure long-term operational reliability. To address these concerns, researchers are developing more stable solvents and additives that inhibit degradation, as well as more efficient process designs that minimize solvent loss. Regulatory frameworks and environmental guidelines also play a role in ensuring that solvent-based systems are deployed in a manner that safeguards human health and ecosystems.

Recent Advances and Innovations

In recent years, significant advancements have been made in improving solvent-based capture technologies. One promising development is the use of biphasic solvents, which separate into two liquid phases upon CO_2 absorption, allowing for more efficient regeneration. This approach reduces the energy required for stripping and enhances overall system performance. Another area of innovation is the development of advanced process configurations, such as absorber intercooling and heat integration techniques, which reduce energy consumption and improve capture efficiency. The use of membrane-assisted solvent processes, where membranes are integrated with solvent absorption systems, is also being explored to achieve hybrid performance benefits. Moreover, artificial intelligence and machine learning are increasingly being applied to optimize process control, monitor solvent condition, and predict system performance. These digital tools enhance reliability, reduce downtime, and enable predictive maintenance, contributing to more efficient and cost-effective operation.

Cost and Commercial Viability

Cost remains one of the most significant barriers to the widespread deployment of solvent-based capture systems. The capital and operating expenses associated with these systems, including solvent procurement, energy consumption, maintenance, and waste management, can be substantial. However, ongoing improvements in solvent formulations, process integration, and scale economies are gradually reducing costs. Government incentives, carbon pricing mechanisms, and international funding initiatives are also helping to close the economic gap. The success of commercial-scale projects such as Petra Nova in the United States and Boundary Dam in Canada demonstrates that solvent-based capture can be economically viable under the right conditions. Continued investment and supportive policy frameworks are essential to accelerate deployment and achieve global climate goals.

Future Outlook and Role in Climate Mitigation

As the world seeks to limit global warming to well below 2°C, solvent-based capture systems are expected to play an increasingly important role. They offer a reliable and proven method for reducing emissions from existing fossil-based infrastructure and hard-to-abate industrial sources. Their flexibility, scalability, and compatibility with retrofit applications make them particularly attractive in regions that depend heavily on coal and natural gas. Looking ahead, solvent-based systems will likely evolve through the integration of novel solvents, advanced process designs, and digital technologies. Their deployment will be supported by the development of CO_2 transport and storage infrastructure, as well as robust regulatory and financial mechanisms. With continued innovation and collaboration, solvent-based capture systems can contribute significantly to global decarbonization efforts and the realization of net-zero emissions targets.

2.1.3 Emerging Technologies in Post-Combustion Capture

The growing urgency to address climate change has pushed forward the development of innovative technologies to reduce carbon dioxide emissions from fossil fuel-based power plants and industrial sources. Among the various carbon capture strategies, post-combustion capture is one of the most applicable, as it can be retrofitted to existing facilities without the need for redesigning the entire process infrastructure. Traditional post-combustion capture methods, primarily based on solvent systems like monoethanolamine (MEA), have proven effective but are energy-intensive and costly. This has driven research into emerging technologies that aim to enhance efficiency, lower costs, and overcome the limitations of conventional methods. These technologies include advanced solvents, solid sorbents, membrane separation, cryogenic processes, enzyme-based capture, and hybrid systems. Each innovation offers unique advantages and challenges, and collectively they represent the future of post-combustion CO_2 capture. Emerging technologies in post-combustion capture represent the next frontier in carbon dioxide mitigation efforts. By overcoming the limitations of conventional solvent-based systems, these innovations promise to enhance the efficiency, scalability, and sustainability of CO_2 capture. Whether through advanced solvents, solid sorbents, membrane systems, cryogenic techniques, or enzyme-based processes, the future of post-combustion capture is marked by diversity and technological synergy. As these technologies mature and are deployed in real-world settings, they will become crucial tools in achieving global climate targets and building a carbon-neutral future.

Advanced Solvents and Solvent Blends

Traditional amine-based solvents have been widely used in post-combustion capture, but they suffer from drawbacks such as high energy consumption for regeneration, solvent degradation, and corrosiveness. In response, researchers have developed advanced solvents and amine blends that exhibit improved thermal stability, higher CO_2 loading capacity, and reduced regeneration energy requirements. For instance, sterically hindered amines prevent the formation of stable carbamates, enabling easier release of CO_2 during regeneration. Tertiary amines such as MDEA and blends incorporating piperazine have shown greater efficiency and lower volatility, making them more sustainable over long-term operations.

Another significant development is the formulation of biphasic solvents. These solvents form two immiscible liquid phases after CO_2 absorption, with one phase being rich in CO_2 and the other lean. This separation allows only the CO_2-rich phase to be sent for regeneration, which reduces the overall energy demand. Similarly, amino acid salts have gained attention for their low environmental impact and high absorption capacity. While these solvent innovations are still under pilot testing, they offer promising routes to cost-effective and efficient CO_2 capture.

Solid Sorbents and Adsorption Technologies

Solid sorbent-based systems offer an alternative to liquid solvent methods by utilizing materials that physically or chemically bind CO_2 onto their surfaces. Materials such as activated carbon, zeolites, metal-organic frameworks (MOFs), and amine-functionalized silicas are among the most widely studied. These solid sorbents provide high surface areas and tunable chemical properties, enabling selective and reversible CO_2 adsorption.

Pressure Swing Adsorption (PSA) and Temperature Swing Adsorption (TSA) are common methods used in these systems, relying on pressure or temperature changes to release the captured CO_2. One key advantage of solid sorbents is their lower energy requirement for regeneration compared to traditional solvents. They also have less corrosion risk and produce fewer environmental pollutants.

However, challenges such as sorbent degradation, reduced capacity at low CO_2 concentrations, and sensitivity to moisture must be addressed before commercial deployment becomes widespread. Research is ongoing to improve sorbent durability and develop scalable systems for real-world applications.

Membrane Separation Technologies

Membrane-based CO_2 separation is another emerging post-combustion capture approach. These systems use selective membranes that allow CO_2 to pass through while retaining other flue gas components such as nitrogen and oxygen. Membranes can be fabricated from polymers, ceramics, or advanced composite materials. They operate on the principle of partial pressure differences across the membrane surface, allowing for continuous CO_2 separation without phase change. One of the primary advantages of membrane systems is their modularity and compact design, which makes them easy to integrate into existing plants. They also have no moving parts, reducing mechanical complexity and maintenance requirements. Additionally, membranes do not require chemical solvents, which avoids problems associated with solvent management and disposal. However, current membrane technologies face limitations in terms of selectivity, permeability, and long-term stability. Advanced materials such as facilitated transport membranes and mixed matrix membranes are under development to address these issues. These next-generation membranes incorporate CO_2 carriers or nanoporous fillers to improve performance and bring them closer to commercial viability.

Cryogenic CO_2 Capture

Cryogenic CO_2 capture involves cooling the flue gas to very low temperatures where CO_2 can be condensed or solidified and separated from the other gas components. This technology is highly selective and results in high-purity CO_2 suitable for compression and storage. It is particularly attractive for applications involving high CO_2 concentrations or where low-temperature utilities are readily available, such as in liquefied natural gas (LNG) facilities. Recent innovations in cryogenic processes aim to improve energy efficiency through advanced heat integration and phase change control. For example, integrated cooling cycles and cascaded refrigeration systems are being designed to reduce the cost and energy penalty of CO_2 liquefaction. Despite these improvements, cryogenic capture remains energy-intensive for low CO_2 concentration flue gases, which is a limitation in typical post-combustion applications. Continued research is focusing on optimizing these systems and combining them with other technologies for enhanced performance.

Enzyme-Based Carbon Capture

Enzyme-based CO_2 capture utilizes naturally occurring or engineered enzymes such as carbonic anhydrase to accelerate the conversion of CO_2 into bicarbonate in aqueous solutions. These biological catalysts offer fast reaction rates, high selectivity, and the ability to operate under mild conditions. When integrated with solvent systems, enzymes can dramatically enhance absorption kinetics, enabling smaller equipment sizes and lower energy consumption. Research efforts are directed toward improving enzyme stability at elevated temperatures and in harsh flue gas environments. Immobilization techniques, such as attaching enzymes to solid supports or encapsulating them in polymers, are being developed to enhance reusability and protect against degradation. Enzyme-based systems are still at the experimental stage but have the potential to revolutionize post-combustion capture by combining the benefits of biotechnology with traditional chemical engineering.

Hybrid and Integrated Systems

Hybrid capture systems combine two or more separation technologies to leverage their individual strengths and compensate for their weaknesses. For example, solvent-membrane hybrid systems use membranes for pre-concentration followed by solvent absorption for final polishing. This configuration can reduce the solvent volume and energy needed for regeneration while increasing overall CO_2 capture efficiency. Another approach involves integrating adsorption and cryogenic technologies, where a solid sorbent is used for initial capture and cryogenic separation is used to purify the CO_2 stream. These integrated systems are being developed with a focus on reducing energy consumption, lowering operating costs, and improving the flexibility of CO_2 capture operations. Moreover, process intensification strategies such as rotating packed beds, microreactors, and 3D-printed absorbers are being explored to enhance mass transfer and reaction rates in compact units. Such designs aim to minimize the physical footprint of capture systems and improve performance in challenging industrial environments.

Digitalization and Smart Control

Emerging post-combustion capture technologies are increasingly incorporating digital tools such as artificial intelligence (AI), machine learning, and real-time monitoring to optimize performance. These tools can be used to model complex processes, predict equipment behavior, and enhance system reliability. For example, AI algorithms can be trained on operational data to optimize solvent circulation rates, anticipate solvent degradation, or detect early signs of equipment failure. Digital twin technologies, which create virtual models of capture systems, enable operators to simulate different scenarios and optimize configurations before implementation. This reduces commissioning time and supports continuous improvement throughout the lifecycle of the system. The use of smart sensors and data analytics is making post-combustion capture not only more efficient but also more adaptive to varying plant conditions.

Challenges and the Road Ahead

Despite the progress in emerging post-combustion capture technologies, several challenges remain. High initial capital costs, limited field demonstrations, and regulatory uncertainties can hinder the commercialization of these innovations. Scaling up laboratory successes to full-scale operations requires significant investment and collaboration across industry, academia, and government agencies. In addition, the economic competitiveness of these technologies depends heavily on policy support, such as carbon pricing, tax incentives, and funding for research and pilot projects.

Public acceptance, particularly regarding the safety and environmental impact of capture technologies, also plays a critical role in determining their widespread adoption. Nonetheless, the ongoing development of advanced capture technologies offers hope for more efficient, cost-effective, and environmentally friendly solutions. As the world moves toward net-zero targets, emerging post-combustion capture systems are expected to play a vital role in reducing emissions from existing fossil fuel infrastructure and hard-to-abate industrial sectors.

2.2 PRE-COMBUSTION CAPTURE

2.2.1 Gasification and Shift Reactors

Pre-combustion capture is a method of carbon capture and storage (CCS) that involves removing carbon dioxide (CO_2) before the combustion of fuel takes place. This approach is typically used in power plants and industrial facilities where the fuel, such as coal, natural gas, or biomass, is processed to produce a mixture of hydrogen and carbon dioxide. The primary technologies enabling this process are gasification and shift reactors, which work together to convert solid or liquid fuels into a clean, hydrogen-rich gas.

By separating CO_2 from this mixture before it is combusted, pre-combustion capture allows for efficient and concentrated removal of greenhouse gases, offering a promising path toward decarbonizing energy and industrial systems. Gasification and shift reactors form the technological backbone of pre-combustion carbon capture systems. By converting fossil fuels into hydrogen-rich syngas and separating CO_2 before combustion, these technologies offer a high-efficiency route to low-carbon energy production.

While challenges related to cost, complexity, and integration remain, the advantages of concentrated CO_2 streams, hydrogen co-production, and high capture efficiency make pre-combustion capture a compelling option. As the world intensifies efforts to limit global warming and transition to a sustainable energy future, gasification and shift reactors will continue to play a vital role in reducing emissions and enabling a cleaner, hydrogen-powered economy.

Pre-Combustion Capture

Pre-combustion capture operates by converting a carbon-based fuel into a synthetic gas, or syngas, which primarily consists of hydrogen (H_2), carbon monoxide (CO), and smaller amounts of other gases. This syngas is processed in shift reactors, where CO reacts with steam to form CO_2 and additional hydrogen. The CO_2 is then separated from the gas mixture and either compressed for storage or transported via pipelines to long-term geological storage sites. The resulting hydrogen can then be used in gas turbines, fuel cells, or industrial processes, emitting only water vapor when combusted. This sequence makes pre-combustion capture a valuable strategy for reducing emissions in energy-intensive sectors.

Role of Gasification in Pre-Combustion Systems

Gasification is the initial step in the pre-combustion process, wherein carbon-based feedstocks like coal, petroleum residues, or biomass are subjected to high temperatures and controlled amounts of oxygen or steam to produce syngas. Unlike combustion, which fully oxidizes the fuel, gasification involves partial oxidation, breaking down the complex hydrocarbons into simpler gaseous molecules. The process occurs in a gasifier, a high-pressure vessel designed to withstand extreme temperatures and corrosive gases.

The resulting syngas is a mixture primarily composed of CO and H_2, but it also contains impurities such as hydrogen sulfide (H_2S), ammonia, and particulates. These contaminants must be removed through gas cleaning systems before further processing. Gasification offers several advantages, including the ability to handle a variety of low-grade fuels and the production of a high-pressure gas stream, which is conducive to efficient CO_2 separation. Additionally, it enables the co-production of electricity, fuels, and chemicals, making it a versatile technology for integrated energy systems.

Shift Reactors and the Water-Gas Shift Reaction

Following gasification, the syngas enters shift reactors where the water-gas shift (WGS) reaction occurs. This process is critical for increasing the hydrogen yield and converting CO into CO_2, which can then be captured. In the WGS reaction, carbon monoxide reacts with steam over a catalyst to form carbon dioxide and hydrogen, as shown in the equation: $CO + H_2O \rightarrow CO_2 + H_2$. This reaction is exothermic, meaning it releases heat, and typically takes place in two stages: a high-temperature shift followed by a low-temperature shift to maximize conversion efficiency. Catalysts used in the shift reaction, often based on iron or copper, facilitate the conversion at different temperature ranges. The high-temperature shift reactor operates at around 350°C and converts most of the CO, while the low-temperature shift reactor, functioning at around 200°C, ensures the reaction goes nearly to completion. The resulting gas stream, rich in hydrogen and carbon dioxide, is then processed to separate the CO_2. The use of shift reactors is essential in pre-combustion capture, as it transforms the syngas into a form where CO_2 can be efficiently extracted and hydrogen can be utilized as a clean energy source.

CO_2 Separation and Capture Technologies

Once the shift reaction has converted CO into CO_2, the next step is separating CO_2 from the hydrogen-rich gas mixture. The separation is typically performed under high pressure, which simplifies the process compared to post-combustion capture. Several technologies are used for this purpose, including physical solvents like Selexol and Rectisol, which absorb CO_2 and other acid gases from the gas stream. These solvents operate efficiently at high pressures and low temperatures, allowing for the capture of concentrated streams of CO_2.Other methods for CO_2 separation include pressure swing adsorption (PSA), where gases are separated based on their adsorption characteristics at varying pressures, and membrane-based technologies that selectively allow CO_2 or hydrogen to pass through. Each method has its own operational characteristics, with solvent-based systems being the most mature and widely applied in large-scale facilities. The captured CO_2 is then dehydrated, compressed, and either stored in deep geological formations or utilized in industrial applications such as enhanced oil recovery or chemical synthesis.

Hydrogen Utilization and Clean Energy Production

The hydrogen separated from the syngas after CO_2 capture serves as a clean fuel for electricity generation and other applications. When hydrogen is burned or used in fuel cells, it produces only water vapor, making it an ideal energy carrier for a decarbonized economy. In pre-combustion systems, hydrogen can be directly combusted in gas turbines to generate electricity or blended with natural gas for use in existing infrastructure. Moreover, hydrogen can serve as a feedstock in the production of ammonia, methanol, and synthetic fuels, contributing to the decarbonization of various industrial sectors. The integration of pre-combustion capture technologies with hydrogen production supports the broader transition to a hydrogen economy, where clean hydrogen becomes a central element of energy systems. This dual benefit of CO_2 mitigation and hydrogen production positions pre-combustion capture as a strategic technology in climate change mitigation.

Advantages of Pre-Combustion Capture

Pre-combustion capture offers several advantages over post-combustion methods. Firstly, the high pressure and concentration of CO_2 in the syngas stream make it easier and more energy-efficient to capture and compress CO_2. Secondly, the production of hydrogen as a by-product allows for cleaner energy production and supports the development of low-carbon fuels. Thirdly, pre-combustion systems can be integrated with combined cycle power plants, improving overall efficiency and enabling cogeneration of electricity and useful products. The technology is particularly suited for new-build facilities such as Integrated Gasification Combined Cycle (IGCC) plants, which are designed from the ground up to incorporate gasification, shift reactors, and CO_2 capture. Pre-combustion capture also aligns well with carbon capture, utilization, and storage (CCUS) strategies, allowing for flexible applications across energy and industrial sectors.

Challenges and Limitations

Despite its advantages, pre-combustion capture faces several challenges. The complexity and cost of gasification and shift reactor systems make them less economically attractive compared to traditional combustion technologies. High capital investment, long construction timelines, and operational risks can deter deployment, especially in regions with limited financial and technical resources. Additionally, retrofitting existing power plants with pre-combustion capture is often impractical due to the fundamental redesign required. Another challenge is the competition with emerging renewable energy sources, which are rapidly gaining market share due to falling costs and favorable policies. However, pre-combustion capture still holds value in hard-to-decarbonize sectors such as heavy industry, aviation fuels, and long-duration energy storage, where direct electrification is difficult or impossible.

2.2.2 Separation and Hydrogen Production

Pre-combustion carbon capture is a crucial component in the broader carbon capture and storage (CCS) framework that targets significant reductions in carbon dioxide emissions from industrial and energy-related sources. Among the various methods of carbon capture, pre-combustion techniques stand out for their capacity to produce high-purity hydrogen while simultaneously enabling the separation of carbon dioxide before combustion occurs. This dual outcome is particularly valuable in decarbonizing the energy and industrial sectors that are heavily dependent on fossil fuels. Pre-combustion capture involves transforming carbon-based fuels into a synthesis gas composed mainly of hydrogen and carbon monoxide, followed by the conversion of carbon monoxide into carbon dioxide and hydrogen. The separation of these gases allows for the storage of carbon dioxide and the utilization of hydrogen as a clean energy carrier. Separation and hydrogen production through pre-combustion capture represent a significant opportunity to reduce greenhouse gas emissions while enabling the generation of clean energy. By transforming carbon-based fuels into hydrogen and carbon dioxide before combustion, this approach allows for efficient CO_2 removal and the utilization of hydrogen as a versatile energy carrier. Although challenges remain in terms of cost and infrastructure, the environmental benefits and potential for integration into a broader hydrogen economy make pre-combustion capture an important tool in the global effort to mitigate climate change. With continued innovation and supportive policy frameworks, this technology can contribute meaningfully to achieving net-zero emissions in the decades ahead.

Principles of Pre-Combustion Capture

The foundation of pre-combustion carbon capture lies in the chemical transformation of fossil fuels or biomass before they are combusted. Unlike post-combustion systems that capture carbon dioxide from flue gases after burning, pre-combustion capture removes the carbon content beforehand. This is done by gasifying the fuel to produce synthesis gas, also known as syngas, which contains hydrogen and carbon monoxide. Through the water-gas shift reaction, carbon monoxide is reacted with steam to produce carbon dioxide and additional hydrogen. The carbon dioxide is then separated from the gas stream, leaving behind hydrogen which can be used as a clean fuel in power generation or industrial applications. The ability to isolate carbon dioxide in a concentrated form under high pressure makes pre-combustion capture more efficient and economically viable in specific contexts, especially in integrated systems like gasification-based power plants.

Production of Synthesis Gas as a First Step

The initial phase in pre-combustion carbon capture is the production of syngas via processes like gasification or reforming. Gasification is commonly used for solid fuels such as coal or biomass, while steam methane reforming is applied for natural gas. In gasification, a carbon-rich material is exposed to a controlled amount of oxygen or steam at high temperatures, resulting in the production of a gas mixture primarily composed of carbon monoxide and hydrogen. In steam methane reforming, methane reacts with steam to yield carbon monoxide and hydrogen. These processes create a gas stream that is amenable to further treatment through the water-gas shift reaction. The composition of syngas can be adjusted to maximize hydrogen production, making it an ideal precursor for both clean energy applications and carbon dioxide removal.

Water-Gas Shift Reaction and Its Role

After the syngas is generated, it is subjected to the water-gas shift reaction to enhance hydrogen yield and convert carbon monoxide into carbon dioxide. The reaction, which occurs when carbon monoxide reacts with steam, results in carbon dioxide and hydrogen. This process is typically carried out in two stages to optimize efficiency. In the high-temperature shift stage, most of the carbon monoxide is converted at elevated temperatures using iron-based catalysts. This is followed by a low-temperature shift stage using copper-based catalysts, which enables further conversion and maximizes hydrogen production. The result is a gas stream enriched in hydrogen and carbon dioxide, which can then be separated efficiently. This step is critical because it enables the pre-combustion process to deliver both a high-purity hydrogen stream and a concentrated carbon dioxide stream suitable for compression and storage.

Gas Separation Techniques for CO_2 and Hydrogen

The separation of carbon dioxide and hydrogen from the shifted syngas mixture is a key component in the pre-combustion capture process. Since the gas mixture is typically at high pressure, it is favorable for efficient separation. One common method for carbon dioxide removal is physical absorption using solvents such as Selexol or Rectisol. These solvents operate effectively at high pressures and low temperatures, enabling the absorption of carbon dioxide and other acid gases from the syngas. Once absorbed, the carbon dioxide can be released by reducing the pressure or increasing the temperature, after which it can be compressed and stored underground. Hydrogen, which is less soluble in these solvents, remains in the gas phase and is thus effectively separated.

Another separation technique is pressure swing adsorption, which involves passing the gas mixture through beds of adsorbent material that preferentially capture carbon dioxide or other impurities at high pressure. By cycling the pressure, the adsorbed gases can be released, and the adsorbent material regenerated. This method is particularly effective for producing high-purity hydrogen from a gas stream. Membrane separation technologies are also emerging, utilizing materials that allow selective permeation of hydrogen or carbon dioxide. These membranes offer compact and potentially lower-cost alternatives, although they are still under development for large-scale applications.

Hydrogen as a Clean Energy Vector

The hydrogen obtained from pre-combustion processes plays a pivotal role in the transition to a low-carbon energy system. When hydrogen is combusted or used in fuel cells, it produces only water as a by-product, eliminating the emission of carbon dioxide. This makes it an ideal fuel for sectors such as transportation, power generation, and industrial heating. In integrated gasification combined cycle (IGCC) power plants with carbon capture, hydrogen can be used to drive turbines for electricity generation with minimal environmental impact. Additionally, hydrogen can be stored and transported for use in various downstream applications, contributing to the resilience and flexibility of energy systems.

The role of hydrogen extends beyond energy. It is a critical feedstock in the chemical industry, particularly in the production of ammonia and methanol. By replacing hydrogen produced from unabated fossil fuels with hydrogen from pre-combustion capture systems, the chemical industry can significantly reduce its carbon footprint. Moreover, hydrogen can be injected into existing natural gas grids to lower the carbon intensity of heating systems, or blended with natural gas in turbines to reduce emissions from power plants.

Environmental and Economic Considerations

Pre-combustion capture offers environmental benefits by enabling deep reductions in carbon dioxide emissions. Since the carbon dioxide is captured in a concentrated and pressurized form, the energy required for its compression and transport is lower compared to post-combustion methods. This improves the overall efficiency and cost-effectiveness of carbon capture and storage. The production of hydrogen as a valuable by-product adds an economic incentive, particularly in regions with supportive policies for clean hydrogen deployment. Furthermore, pre-combustion systems can be integrated with renewable energy sources, such as biomass gasification, to achieve even greater emissions reductions.

However, the technology also poses economic and technical challenges. The capital costs associated with building gasifiers, shift reactors, and separation units are substantial. The infrastructure required for hydrogen distribution and carbon dioxide storage must also be developed in parallel. These factors can make pre-combustion capture less attractive in regions where natural gas prices are low or where renewable electricity is rapidly becoming more affordable. Moreover, retrofitting existing power plants with pre-combustion systems is often impractical, limiting the application of this technology primarily to new-build projects.

2.2.3 Advantages and Challenges of Pre-Combustion Capture

Pre-combustion carbon capture is a key technological pathway within the broader framework of carbon capture and storage (CCS) aimed at reducing greenhouse gas emissions, particularly carbon dioxide (CO_2). In this method, carbon is separated from fuels before combustion through a series of chemical reactions that transform the fuel into a synthesis gas consisting mainly of hydrogen and carbon monoxide. This syngas is then subjected to the water-gas shift reaction, which converts carbon monoxide into carbon dioxide and additional hydrogen. The carbon dioxide is captured and stored, while hydrogen is utilized as a clean energy carrier. Pre-combustion carbon capture is especially relevant in integrated gasification combined cycle (IGCC) power plants and hydrogen production facilities.

It offers significant advantages in terms of efficiency and effectiveness in capturing CO_2. However, the technology also presents several challenges that limit its widespread adoption. Pre-combustion carbon capture represents a powerful tool for reducing carbon emissions while enabling the production of clean hydrogen energy. Its advantages include high capture efficiency, production of high-purity hydrogen, and suitability for integration with advanced power systems like IGCC. However, it also faces significant hurdles, including high costs, technical complexity, and limited retrofit potential. Addressing these challenges through innovation, policy support, and demonstration projects will be essential for unlocking the full potential of pre-combustion capture. As the world moves toward deep decarbonization, this technology could play an important role in shaping a cleaner, more sustainable energy future.

High Efficiency of CO_2 Capture

One of the most compelling advantages of pre-combustion capture is its high efficiency in removing carbon dioxide from the process stream. Because the separation occurs before combustion, the CO_2 is concentrated and pressurized, making it easier and less energy-intensive to capture. Unlike post-combustion processes, which deal with dilute CO_2 in flue gases at near atmospheric pressure, pre-combustion operates at higher pressures, allowing the use of physical solvents such as Selexol and Rectisol. These solvents absorb CO_2 more efficiently under such conditions, thereby improving the overall energy balance of the system. As a result, pre-combustion systems can achieve capture efficiencies exceeding 90 percent, making them highly effective for emissions reduction.

Production of Clean Hydrogen Fuel

Pre-combustion processes have the unique advantage of simultaneously producing hydrogen, a clean and versatile fuel. After the water-gas shift reaction, the gas stream contains a high proportion of hydrogen, which can be separated and used in power generation, transportation, and industrial applications. This dual benefit—emissions reduction through CO_2 capture and energy supply via hydrogen—makes pre-combustion capture particularly attractive in a future hydrogen economy. Hydrogen produced in this manner can be burned in turbines to generate electricity or used in fuel cells, emitting only water as a by-product. Thus, pre-combustion capture aligns well with long-term strategies for decarbonizing multiple sectors of the economy.

Suitability for Integration with IGCC Plants

Pre-combustion capture is especially suited to integration with integrated gasification combined cycle (IGCC) power plants. These facilities convert coal, biomass, or other feedstocks into syngas, which is then processed to remove CO_2 before the hydrogen-rich gas is combusted in a gas turbine. The heat from this process is used to generate additional electricity via a steam turbine, significantly improving overall efficiency. The IGCC setup not only allows for the efficient removal of CO_2 but also offers the flexibility to use a range of feedstocks, making it a robust solution for regions with diverse energy resources. Furthermore, the synergy between gasification, hydrogen production, and carbon capture enables deep decarbonization of power generation with fewer modifications to existing infrastructure.

High Purity and Concentration of Captured CO_2

Another advantage of pre-combustion capture is that it yields a high-purity CO_2 stream, which is well-suited for transport and long-term geological storage. This reduces the cost and complexity associated with CO_2 purification and compression. High-purity CO_2 can also be used for enhanced oil recovery (EOR) and other industrial applications, creating additional economic incentives for capture projects. In contrast, post-combustion systems often require significant energy input to separate and purify CO_2 from a mixture of gases, which can offset the environmental benefits of capture. The simplicity of the gas stream in pre-combustion setups translates into more cost-effective and technically straightforward operations.

Challenges in Retrofitting Existing Facilities

Despite its many advantages, pre-combustion capture is difficult to retrofit into existing power plants and industrial facilities. The process requires a complete redesign of the combustion system, including the installation of gasifiers, shift reactors, and advanced separation units. This makes it economically viable only for new-build projects, limiting its immediate applicability in reducing emissions from the current fleet of power plants and factories. In contrast, post-combustion systems can often be added to existing infrastructure with relatively fewer modifications. The high capital costs and long lead times associated with pre-combustion installations make them less attractive in the short term, particularly in regions with aging energy infrastructure.

High Capital and Operational Costs

The deployment of pre-combustion capture systems involves significant capital investment. Gasification equipment, water-gas shift reactors, and CO_2 separation units are all complex and expensive technologies. Additionally, the integration of these components into a cohesive and efficient system requires sophisticated engineering and control systems.

Operational costs are also high, due in part to the need for high-pressure operation and precise temperature control throughout the process. These economic barriers can discourage investment, especially in markets where carbon pricing is weak or inconsistent. Without strong policy support and financial incentives, the economic case for pre-combustion capture remains limited.

Technical Complexity and Maintenance

The technical complexity of pre-combustion systems presents additional challenges. The gasification process must be carefully managed to ensure consistent syngas quality, while the water-gas shift reaction requires precise control to optimize hydrogen and CO_2 yields. Equipment must operate under high pressures and temperatures, increasing the risk of wear, corrosion, and failure. Maintenance requirements are therefore more demanding than in conventional power plants, and system downtime can be costly. The need for specialized knowledge and skilled operators further complicates the widespread adoption of pre-combustion technologies, especially in regions with limited technical capacity.

Limited Commercial Deployment

While the theoretical and pilot-scale benefits of pre-combustion capture are well established, its commercial deployment has been limited. Only a few large-scale projects, such as the Kemper County Energy Facility and the Hydrogen Energy California project, have attempted to implement pre-combustion capture on a commercial scale, and many of these have faced delays, cost overruns, or cancellation. The lack of successful large-scale demonstrations undermines investor confidence and slows down the development of supply chains and regulatory frameworks needed to support the technology. In contrast, post-combustion capture and renewable energy technologies have seen more rapid deployment and cost reductions in recent years, placing additional competitive pressure on pre-combustion systems.

Water Usage and Environmental Impact

Pre-combustion systems often require significant amounts of water, particularly for the gasification and water-gas shift reactions. This can pose environmental concerns in water-scarce regions and complicate the siting of new facilities. Additionally, the production and disposal of solvent materials used in CO_2 separation can have environmental implications. The management of by-products from gasification, such as slag and tar, also requires careful handling to prevent pollution. While the overall goal of pre-combustion capture is to reduce greenhouse gas emissions, the environmental trade-offs associated with its implementation must be carefully managed to ensure net positive outcomes.

Future Prospects and Pathways

Despite its challenges, pre-combustion carbon capture holds significant promise for the future, especially as part of a diversified strategy to achieve net-zero emissions. Advances in gasification technology, materials science, and system integration could help reduce costs and improve performance. Innovations in hydrogen storage and utilization may also enhance the value proposition of pre-combustion systems. Government policies that support clean hydrogen and impose stricter emissions limits will be crucial in driving demand and investment. As global efforts to decarbonize intensify, pre-combustion capture could find increasing relevance in sectors that are difficult to electrify, such as steel and cement production.

2.3 OXY-FUEL COMBUSTION

2.3.1 Oxy-Fuel Combustion Process

Oxy-fuel combustion is the process of burning a fuel using pure oxygen, or a mixture of oxygen and recirculated flue gas, instead of air. Since the nitrogen component of air is not heated, fuel consumption is reduced, and higher flame temperatures are possible. Historically, the primary use of oxy-fuel combustion has been in welding and cutting of metals, especially steel, since oxy-fuel allows for higher flame temperatures than can be achieved with an air-fuel flame. It has also received a lot of attention in recent decades as a potential carbon capture and storage technology. There is currently research being done in firing fossil fuel power plants with an oxygen-enriched gas mix instead of air. Almost all of the nitrogen is removed from input air, yielding a stream that is approximately 95% oxygen. Firing with pure oxygen would result in too high a flame temperature, so the mixture is diluted by mixing with recycled flue gas, or staged combustion. The recycled flue gas can also be used to carry fuel into the boiler and ensure adequate convective heat transfer to all boiler areas. Oxy-fuel combustion produces approximately 75% less flue gas than air fueled combustion and produces exhaust consisting primarily of CO_2 and H_2O.

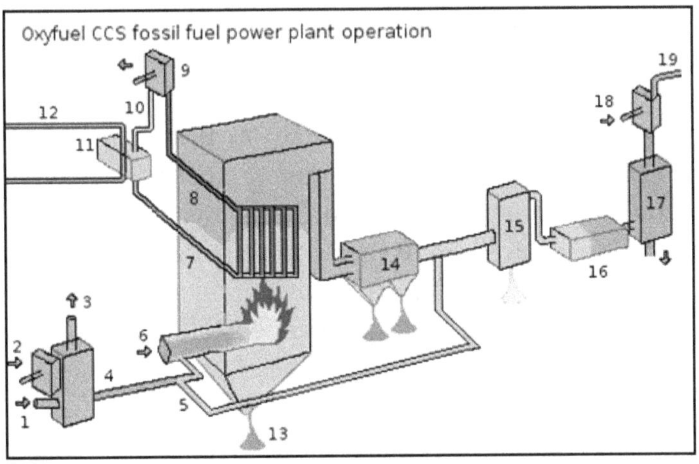

Fig: Oxyfuel CCS Power Plant Operation

Economy and Efficiency

The justification for using oxy-fuel is to produce a CO_2 rich flue gas ready for sequestration. Oxy-fuel combustion has significant advantages over traditional air-fired plants. Among these are:

- The mass and volume of the flue gas are reduced by approximately 75%.
- Because the flue gas volume is reduced, less heat is lost in the flue gas.
- The size of the flue gas treatment equipment can be reduced by 75%.
- The flue gas is primarily CO_2, suitable for sequestration.
- The concentration of pollutants in the flue gas is higher, making separation easier.
- Most of the flue gases are condensable; this makes compression separation possible.
- Heat of condensation can be captured and reused rather than lost in the flue gas.
- Because nitrogen from air is absent, nitrogen oxide production is greatly reduced.
- If the fuel contains sulfur, sulfuric acid can possibly be recovered instead of being released as a dangerous environmental pollutant or "lost" in flue gas desulfurization.

Economically speaking this method costs more than a traditional air-fired plant. The main problem has been separating oxygen from the air. This process requires much energy, nearly 15% of production by a coal-fired power station can be consumed for this process. However, a new technology which is not yet practical called chemical looping combustion can be used to reduce this cost. In chemical looping combustion, the oxygen required to burn the coal is produced internally by oxidation and reduction reactions, as opposed to using more expensive methods of generating oxygen by separating it from air. At present in the absence of any need to reduce CO_2 emissions, oxy-fuel is not competitive. However, oxy-fuel is a viable alternative to removing CO_2 from the flue gas from a conventional air-fired fossil fuel plant. However, an oxygen concentrator might be able to help, as it simply removes nitrogen. In industries other than power generation, oxy-fuel combustion can be competitive due to higher sensible heat availability. Oxy-fuel combustion is common in various aspects of metal production.

The glass industry has been converting to oxy-fuel since the early 1990s because glass furnaces require a temperature of approximately 1500 degrees C, which is not economically attainable at adiabatic flame temperatures for air-fuel combustion unless heat is regenerated between the flue stream and the incoming air stream. Developed in the mid-19th century, glass furnace regenerators are large and expensive high temperature brick ducts filled with brick arranged in a checkerboard pattern to capture heat as flue gas exits the furnace. When the flue duct is thoroughly heated, air flow is reversed and the flue duct becomes the air inlet, releasing its heat into the incoming air, and allowing for higher furnace temperatures than can be attained with air-fuel only. Two sets of regenerative flue ducts allowed for the air flow to be reversed at regular intervals, and thus maintain a high temperature in the incoming air. By allowing new furnaces to be built without the expense of regenerators, and especially with the added benefit of nitrogen oxide reduction, which allows glass plants to meet emission restrictions, oxy-fuel is cost effective without the need to reduce CO_2 emissions. Oxy-fuel combustion also reduces CO_2 release at the glass plant location, although this may be offset by CO_2 production due to electric power generation which is necessary to produce oxygen for the combustion process. Oxy-fuel combustion may also be cost effective in the incineration of low BTU value hazardous waste fuels. It is often combined with staged combustion for nitrogen oxide reduction, since pure oxygen can stabilize combustion characteristics of a flame.

Pilot Plants

There are pilot plants undergoing initial proof-of-concept testing to evaluate the technologies for scaling up to commercial plants, including

- Callide A Power Station in Queensland, Australia
- Schwarze Pumpe Power Station in Spremberg, Germany
- CIUDEN in Cubillos del Sil, Spain
- NET Power Demonstration Facility in La Porte, Texas, USA

White Rose Plant

One case study of oxy-fuel combustion is the attempted White Rose plant in North Yorkshire, United Kingdom. The planned project was an oxy-fuel power plant coupled with air separation to capture two million tons of carbon dioxide per year. The carbon dioxide would then be delivered by pipeline to be sequestered in a saline aquifer beneath the North Sea. However, in late 2015 and early 2016, following withdrawal of funding by the Drax Group and the U.K. government, construction was halted. The unforeseen loss of the funding from the UK government's CCS Commercialisation Programme, along with decreased subsidies for renewable energy, left the White Rose Plant with insufficient funds to continue development.

Environmental Impact

One of the major environmental impacts of burning fossil fuels is the release of CO_2, which contributes to climate change. Because oxyfuel combustion results in flue gas that already has a high concentration of CO_2, it makes it easier to purify and store the CO_2 rather than releasing it to the atmosphere. Many fossil fuels, such as coal and oil shale, produce ash as a result of combustion. This ash also needs to be disposed of, which may impact the environment. So far studies indicate that, in general, oxyfuel combustion does not significantly affect the composition of ash produced. Measurements have shown similar mineral and heavy metal concentrations regardless of whether an air or oxyfuel environment was used. However, one notable exception is that oxyfuel ashes often have lower concentrations of calcium oxide or calcium hydroxide (free lime). Free lime forms when carbonate minerals in fuels like coal and oil shale decompose at the high temperatures occurring during combustion (calcination). Calcination is an equilibrium reaction and a higher partial pressure of CO_2 shifts the equilibrium in favor of CaCO3 and MgCO3 respectively. Free lime is reactive and can potentially affect the environment, for instance by increasing the alkalinity of the ash.

Because oxyfuel combustion takes place in a CO_2-rich atmosphere, decomposition is reduced and the ash generally contains less free lime. Flue gas desulfurization is usually employed to increase the pH of flue gases or their product when reacting with atmospheric moisture (acid rain). Besides sulfur and its oxides, another potential acid rain component is formed from nitric and nitrous oxide interacting with water - eliminating nitrogen from combustion reduces this factor altogether.

2.3.2 Advantages and Challenges of Oxy-Fuel Technology

Oxy-fuel combustion technology represents a promising solution in the field of carbon capture and storage (CCS) by facilitating the separation and capture of carbon dioxide emissions during the combustion process. This approach involves burning fuel in a mixture of pure oxygen and recycled carbon dioxide or steam, rather than air. As a result, the flue gas produced is primarily composed of carbon dioxide and water vapor, which simplifies the process of capturing and compressing CO_2 for storage. Oxy-fuel technology is applicable to both new and retrofitted power plants and has been considered a strong candidate for large-scale deployment in carbon-intensive industries. Despite its many advantages, the widespread adoption of oxy-fuel combustion faces several challenges, including high costs, oxygen supply issues, and technical limitations.

Oxy-fuel combustion presents a powerful option in the suite of carbon capture technologies, offering high carbon dioxide capture rates, reduced nitrogen oxide emissions, and compatibility with existing infrastructure. Its ability to produce high-purity CO_2 makes it well-suited for both storage and industrial applications such as enhanced oil recovery. However, the technology also faces formidable challenges, including high costs, technical complexity, and limited deployment experience. Addressing these challenges will require coordinated efforts in research, policy, and investment. If these barriers can be overcome, oxy-fuel combustion could become a key enabler of deep decarbonization across multiple sectors, contributing meaningfully to the global goal of limiting climate change.

Enhanced Carbon Dioxide Capture Efficiency

One of the most significant advantages of oxy-fuel combustion is its ability to produce a flue gas with a high concentration of carbon dioxide. When air is replaced with pure oxygen during combustion, the nitrogen component that typically makes up around 78 percent of air is eliminated. This results in a cleaner flue gas mixture, consisting mostly of CO_2 and water vapor. Once the water vapor is condensed, a relatively pure stream of CO_2 remains, which can be easily captured and compressed for storage or utilization. This simplifies the carbon capture process compared to post-combustion methods that must separate CO_2 from nitrogen-diluted flue gases. In this way, oxy-fuel technology supports more effective and lower-energy carbon capture, offering a potential capture efficiency of up to 90 percent or more.

Improved Combustion Efficiency and Flame Temperature Control

Oxy-fuel combustion allows for better control over the flame temperature within the boiler or furnace. Combustion with pure oxygen leads to a significantly higher flame temperature than with air; however, this can be moderated by recycling part of the flue gas into the combustion chamber. This process helps maintain optimal temperature levels, improving thermal efficiency while minimizing the risk of equipment damage due to excessive heat. Moreover, the precise control over combustion conditions also enhances the overall fuel utilization and energy output. This is particularly beneficial in industrial processes requiring high thermal intensity, such as cement and steel production.

Reduction of Nitrogen Oxide Emissions

The absence of nitrogen in the combustion environment of oxy-fuel systems leads to a substantial reduction in nitrogen oxide (NO_x) emissions. Nitrogen oxides are harmful pollutants formed when nitrogen and oxygen in the air react at high temperatures during conventional combustion. These compounds contribute to smog formation, acid rain, and respiratory problems. Since oxy-fuel combustion eliminates the nitrogen source from the oxidizer, the formation of NO_x is drastically reduced. This environmental benefit aligns with stricter air quality regulations and supports efforts to mitigate multiple types of emissions beyond just carbon dioxide.

Compatibility with Existing Infrastructure

Oxy-fuel technology can be integrated into both new power plants and existing facilities through retrofitting. Retrofitting conventional power plants to incorporate oxy-fuel combustion systems involves modifying the combustion chamber and flue gas handling equipment but does not necessarily require a complete redesign of the plant. This compatibility offers an opportunity to decarbonize existing energy infrastructure without total replacement, providing a practical and less disruptive transition toward cleaner operations. Moreover, the technology can be adapted to different fuel types, including coal, natural gas, and biomass, offering flexibility for various energy systems and geographic regions.

Facilitation of CO_2 Purity for Enhanced Oil Recovery and Storage

The high-purity carbon dioxide produced from oxy-fuel combustion is well-suited for enhanced oil recovery (EOR) operations and long-term geological storage. In EOR, CO_2 is injected into mature oil reservoirs to increase extraction rates, creating a potential revenue stream that can offset the costs of capture. High-purity CO_2 also reduces the need for extensive purification before storage, lowering overall operational expenses and simplifying the logistics of pipeline transport. These features make oxy-fuel combustion an attractive option for industries looking to integrate carbon capture into economically viable business models.

High Energy Penalty for Oxygen Production

Despite its technical advantages, oxy-fuel combustion faces significant energy penalties, particularly in the production of pure oxygen. The separation of oxygen from air is typically achieved through cryogenic distillation or advanced membrane technologies, both of which are energy-intensive processes. The energy required for oxygen generation can reduce the net efficiency of the overall power or industrial system, thereby offsetting some of the emissions reductions achieved through carbon capture. This additional energy demand increases operating costs and limits the economic competitiveness of oxy-fuel technology, especially in comparison to less energy-intensive alternatives like post-combustion capture.

Increased Capital and Operational Costs

Implementing oxy-fuel combustion systems involves considerable capital investment. Modifications to boilers, turbines, heat exchangers, and flue gas recycling systems are necessary to handle the high temperatures and corrosive environment created by pure oxygen combustion. Additionally, the integration of CO_2 capture, compression, and transport systems further adds to the infrastructure and equipment requirements. Operational costs also rise due to the continuous need for oxygen supply and the maintenance of complex gas processing equipment. These financial barriers present a major challenge for widespread commercial deployment, particularly in regions with limited access to financing or low carbon pricing incentives.

Technical Challenges in Material Durability and System Integration

Oxy-fuel combustion introduces higher combustion temperatures and more reactive gas compositions, which can place significant stress on system materials.

Boiler tubes, heat exchangers, and turbine components must be made of materials capable of withstanding extreme conditions without degradation. This increases the cost and complexity of component design and maintenance. Furthermore, the need to recycle large volumes of flue gas to control temperatures adds complexity to system integration and process control. Ensuring stable operation under these conditions requires advanced monitoring systems and highly skilled operators, adding to the technical barriers of deployment.

Limited Large-Scale Demonstration Projects

Although the theoretical benefits of oxy-fuel combustion are well documented, its real-world deployment at large scale has been limited. Most existing oxy-fuel systems are pilot or demonstration projects, and few have progressed to commercial-scale operations. The absence of widespread implementation limits opportunities for economies of scale, technological learning, and cost reductions. As a result, the knowledge base around long-term system performance, maintenance needs, and reliability remains incomplete. This lack of field experience can hinder investor confidence and policy support, further slowing the pace of development.

Environmental and Safety Considerations

Operating with pure oxygen increases the risk of fire and explosion compared to conventional combustion systems. Special precautions must be taken to prevent oxygen leaks and manage combustion temperatures safely. Additionally, while oxy-fuel combustion reduces CO_2 and NO_x emissions, it can increase concentrations of other pollutants such as sulfur dioxide and mercury if the fuel contains these elements. Effective emission control systems must therefore be integrated to manage the full range of potential pollutants. These environmental and safety considerations require robust regulatory oversight and operational discipline to ensure responsible deployment.

2.4 DIRECT AIR CAPTURE (DAC)

2.4.1 Technologies for Extracting CO_2 from Ambient Air

The growing urgency to mitigate climate change has driven the development of technologies that can actively remove carbon dioxide (CO_2) from the atmosphere. Among these, Direct Air Capture (DAC) stands out as one of the most promising approaches. Unlike traditional carbon capture and storage (CCS) methods, which focus on capturing CO_2 from industrial or power plant emissions, DAC aims to extract CO_2 directly from the ambient air. This capability offers a unique advantage: it can be deployed almost anywhere and can help address emissions from sectors that are difficult to decarbonize, such as aviation, shipping, and agriculture.

Direct Air Capture is therefore seen as an essential tool in achieving net-zero and even net-negative emissions targets. Despite its promise, the technology faces a range of challenges, particularly in terms of energy use, cost, and scalability. Direct Air Capture is an innovative and increasingly vital technology in the fight against climate change. By extracting carbon dioxide directly from the atmosphere, it offers a pathway to remove emissions that are otherwise hard to eliminate. Though currently limited by high costs and energy demands, ongoing technological advances and supportive policies are likely to enhance its viability. Its potential to contribute to negative emissions, especially when linked with permanent storage, positions DAC as a cornerstone of long-term climate strategies. As the world moves toward more ambitious climate goals, the role of DAC will expand, offering hope for balancing the carbon equation and preserving a stable climate for future generations.

Principle of Direct Air Capture

Direct Air Capture involves the use of chemical processes to separate CO_2 from the atmosphere. Since the concentration of CO_2 in ambient air is relatively low, about 420 parts per million, DAC systems must be designed to handle large volumes of air efficiently. The captured CO_2 can either be permanently stored in geological formations or utilized in the production of fuels, building materials, or other products. The key feature that distinguishes DAC from point-source carbon capture is its flexibility in location and its ability to address diffuse emissions. In essence, DAC can complement emissions reductions by actively removing historical CO_2 from the atmosphere, making it an important strategy for achieving climate goals beyond carbon neutrality.

Technologies Used in Direct Air Capture

There are two primary approaches used in DAC systems: liquid solvent-based and solid sorbent-based technologies. In liquid solvent systems, large fans draw air through a contactor where it reacts with a liquid solution, typically containing hydroxide compounds that chemically bind with CO_2. The resulting carbonate solution is then processed through a series of chemical reactions and thermal steps to release pure CO_2 and regenerate the solvent. In solid sorbent systems, air passes over a bed of solid materials, such as amine-functionalized resins, that selectively adsorb CO_2. The captured CO_2 is then released through temperature or vacuum swings, allowing the sorbent to be reused. Each approach has its own advantages and trade-offs in terms of energy efficiency, capture capacity, and operational complexity.

Energy Requirements and Efficiency Considerations

One of the main criticisms of DAC is its high energy demand. Capturing CO_2 from the atmosphere requires significantly more energy than capturing it from concentrated sources like power plant flue gases. This is primarily because the low concentration of CO_2 in air necessitates the processing of vast volumes of air and the use of powerful fans, compressors, and heat systems. Liquid solvent systems often require high-grade heat to release the captured CO_2, while solid sorbent systems may be optimized to use low-temperature heat or electricity. The energy source used in DAC systems greatly affects their overall carbon balance. To be truly carbon-negative, DAC must be powered by renewable energy; otherwise, the emissions from fossil-fueled energy inputs could negate the benefits of CO_2 removal.

Economic Challenges and Cost Considerations

The cost of Direct Air Capture remains a significant barrier to its widespread deployment. Current estimates suggest that DAC costs range from $600 to $1,200 per ton of CO_2 removed, although companies such as Climeworks and Carbon Engineering are working to reduce these figures. The high cost is due to a combination of factors, including capital investment, energy consumption, material regeneration, and maintenance.

However, as with many emerging technologies, these costs are expected to decline with increased deployment, technological improvements, and economies of scale. Supportive policies, such as carbon pricing, subsidies, and investment in research and development, will play a crucial role in accelerating cost reductions and making DAC commercially viable.

Geographical and Infrastructural Flexibility

One of the major strengths of DAC is its ability to be sited independently of emission sources. This means that DAC facilities can be strategically located near renewable energy sources or near suitable geological storage sites for CO_2. This flexibility reduces the need for extensive CO_2 transport infrastructure and allows DAC to be deployed in regions with limited industrial emissions. Additionally, DAC facilities can be modular and scalable, making them suitable for gradual implementation and integration into broader carbon management strategies. This geographic independence also offers opportunities for countries or regions with low emissions to contribute to global CO_2 removal efforts.

Applications and Integration into Circular Carbon Economy

Beyond storage, the CO_2 captured through DAC can be used in various industrial applications. One promising use is in the production of synthetic fuels, where CO_2 is combined with green hydrogen to create carbon-neutral gasoline or jet fuel. Captured CO_2 can also be used in the food and beverage industry, in enhanced oil recovery, and in the manufacture of building materials like concrete.

These applications support the development of a circular carbon economy, where CO_2 is recycled and reused rather than emitted into the atmosphere. However, long-term climate benefits depend on the permanence of CO_2 removal, and not all utilization pathways offer permanent storage. Therefore, the choice of application must align with broader decarbonization objectives.

Technological Progress and Innovation Landscape

The field of DAC is rapidly evolving, with several startups, research institutions, and government programs driving innovation. Companies like Climeworks in Switzerland and Carbon Engineering in Canada have already developed commercial-scale DAC plants, showcasing the feasibility of the technology. Climeworks' Orca and Mammoth facilities in Iceland, for instance, use solid sorbent systems powered by geothermal energy and sequester CO_2 in basalt formations. Carbon Engineering's approach is based on liquid solvent systems and aims to produce synthetic fuels. Meanwhile, research continues into novel materials with higher capture capacities, lower regeneration energy, and improved durability. Emerging solutions, such as metal-organic frameworks (MOFs) and electrochemical DAC systems, hold promise for significantly improving the efficiency and affordability of DAC.

Policy Support and Market Development

The development of Direct Air Capture depends heavily on policy support and the creation of markets for CO_2 removal. As of now, there is limited commercial demand for negative emissions, and few regulatory frameworks offer consistent incentives. Carbon pricing, tax credits, and public procurement programs are essential to support early deployment and drive innovation. In the United States, the 45Q tax credit offers financial support for each ton of CO_2 captured and stored, which has encouraged investment in DAC projects. International initiatives, such as the Mission Innovation Carbon Dioxide Removal initiative, are also providing funding and research collaboration opportunities. As climate commitments become more stringent, the demand for verifiable CO_2 removal is expected to grow, potentially creating a strong market for DAC services.

Environmental and Social Considerations

While DAC is often praised for its climate benefits, it also raises important environmental and social questions. Large-scale DAC deployment could require significant land use, water consumption, and materials, depending on the system design. For example, certain solvent-based systems may use large quantities of water, which could pose sustainability issues in arid regions. Additionally, the siting of DAC plants must consider local communities, land rights, and environmental justice to ensure equitable deployment. Transparency, public engagement, and environmental assessments will be necessary to build trust and avoid unintended consequences as the technology scales up.

Future Prospects and Role in Climate Strategy

Looking ahead, Direct Air Capture is likely to play a critical role in the global effort to achieve net-zero emissions. While emissions reductions must remain the top priority, DAC offers a powerful means to address residual and historical emissions. Its ability to deliver permanent CO_2 removal, especially when combined with geological storage, positions it as a key component in pathways that limit global warming to 1.5°C or 2°C. Realizing this potential will require coordinated efforts across science, policy, industry, and civil society. Technological innovation must continue to improve efficiency and reduce costs, while policy frameworks must create markets and incentives for deployment. As part of a portfolio of solutions, DAC represents not a silver bullet but a necessary tool in the broader climate arsenal.

2.4.2 DAC vs. Point Source Capture: Benefits and Challenges

Direct Air Capture and point source carbon capture are not competing technologies but rather complementary tools in the global fight against climate change. Each has unique advantages and specific roles within the broader decarbonization landscape. Point source capture is efficient and cost-effective for addressing emissions at their origin, particularly in heavy industries. Direct Air Capture offers unmatched flexibility and the ability to remove atmospheric CO_2, making it essential for achieving net-negative emissions. While DAC currently faces higher costs and technological hurdles, it holds significant long-term promise. A balanced and integrated deployment of both technologies, supported by innovation, investment, and public policy, will be crucial for meeting the ambitious climate targets set by the international community.

Point Source Capture

Point source carbon capture is designed to prevent carbon dioxide emissions from entering the atmosphere at the site of emission. It is typically deployed at power plants, cement factories, steel mills, and other heavy industrial facilities. The technology relies on separating CO_2 from flue gases, often using solvent-based chemical absorption methods such as amine scrubbing. Once captured, the CO_2 is either compressed for transportation to a storage site or utilized in various industrial processes. Because the concentration of CO_2 in flue gas is significantly higher than in ambient air, point source capture is generally more energy-efficient compared to DAC. This method has the advantage of tackling emissions from some of the largest sources of greenhouse gases, making it a crucial strategy for reducing emissions in the near term.

Direct Air Capture

Direct Air Capture is a carbon removal technology that draws CO_2 from ambient air using engineered chemical processes. It is not tied to any specific emission source and can therefore be sited independently of industrial operations. DAC systems typically employ either liquid solvents or solid sorbents to bind with atmospheric CO_2, which is then released through heating or pressure changes for storage or utilization. Unlike point source capture, DAC is uniquely capable of removing historic and distributed emissions, including those from sectors that are difficult to decarbonize. Although energy-intensive, DAC provides a means to achieve negative emissions and is increasingly viewed as a critical tool for reaching net-zero targets.

Comparative Efficiency and Energy Use

One of the key differences between DAC and point source capture lies in the concentration of CO_2 they are designed to extract. Flue gases from industrial facilities often contain CO_2 concentrations ranging from 4% to 15%, while ambient air has only around 0.04% CO_2. This disparity makes DAC inherently more energy-intensive because it must process much larger volumes of air to capture the same amount of CO_2. Consequently, DAC systems require more power for fans, compressors, and regeneration of sorbents or solvents. In contrast, point source capture can operate more efficiently due to the relatively high CO_2 concentration and the existing infrastructure at industrial sites that can provide waste heat or power for the capture process. Therefore, from an energy efficiency standpoint, point source capture currently holds a clear advantage.

Cost Implications and Economic Feasibility

Cost remains a significant factor in comparing DAC and point source capture. Point source technologies are generally more mature and cost less per ton of CO_2 captured, typically ranging between \$40 and \$100 per ton, depending on the facility and technology used. DAC, being a newer and more energy-intensive method, currently incurs higher costs, often ranging from \$600 to \$1,200 per ton of CO_2 removed. However, as DAC technology advances and economies of scale are achieved, costs are expected to decline. Furthermore, the economic feasibility of both technologies can be influenced by policy mechanisms such as carbon pricing, tax credits, and direct subsidies. In jurisdictions where such policies are in place, both DAC and point source capture may become more financially attractive, encouraging broader adoption.

Flexibility and Deployment Potential

One of DAC's most notable advantages over point source capture is its geographic flexibility. Because it does not depend on proximity to emission sources, DAC facilities can be located near renewable energy sources or suitable geological storage sites, minimizing the need for long-distance CO_2 transport infrastructure. This contrasts with point source capture, which must be co-located with specific industrial facilities. The location-specific nature of point source systems limits their deployment to particular sectors and regions. DAC's flexibility makes it a valuable option for countries with fewer industrial emissions or those seeking to participate in international carbon markets. Its modular design also allows for incremental deployment, which can ease integration into existing energy systems and carbon reduction plans.

Target Emissions and Environmental Impact

Point source capture is highly effective at preventing new emissions from entering the atmosphere, making it ideal for decarbonizing existing industrial operations. However, it does little to address legacy emissions or CO_2 already present in the atmosphere. DAC, in contrast, provides the capability to remove both current and historic emissions, offering the potential for net-negative emissions. This feature is crucial in scenarios where residual emissions from sectors such as aviation, agriculture, and certain industrial processes are unavoidable. Therefore, DAC complements point source capture by addressing emission sources that are difficult or impossible to eliminate through conventional means. Additionally, if powered by renewable energy, DAC can operate with a low environmental footprint, making it a sustainable long-term solution.

Integration into Climate Policy and Strategy

Both DAC and point source capture play critical roles in national and global climate strategies, particularly in the context of achieving net-zero emissions. Point source capture is essential for mitigating emissions from current fossil fuel-based infrastructure, and its widespread adoption can significantly reduce the carbon intensity of key industries. DAC, on the other hand, is indispensable for offsetting residual emissions and achieving net-negative emissions targets, especially in the latter half of the century. Integrated approaches that combine these technologies can offer a more comprehensive path to decarbonization. Policymakers increasingly recognize the importance of supporting both technologies through robust regulatory frameworks, financial incentives, and investment in research and development.

Technological Maturity and Deployment Challenges

Point source carbon capture technologies are relatively mature and have been demonstrated in multiple industrial applications. Several large-scale CCS projects are currently in operation around the world, particularly in sectors like natural gas processing and power generation. However, widespread adoption still faces barriers such as high capital costs, regulatory uncertainty, and social acceptance.

DAC, in contrast, is still in its early stages of commercialization. Only a handful of operational DAC facilities exist today, and most are relatively small in scale. The scalability of DAC remains one of its greatest challenges, particularly in terms of infrastructure, supply chains, and energy integration. Continued innovation and pilot projects are necessary to improve the efficiency and cost-effectiveness of DAC systems before large-scale deployment becomes feasible.

Public Perception and Social Acceptance

Social and public acceptance is another area where differences between DAC and point source capture are evident. Point source CCS projects have occasionally faced public opposition due to concerns about CO_2 storage safety, environmental risks, and local impacts. In some cases, proposed projects have been delayed or canceled due to resistance from communities and environmental groups. DAC, being less associated with fossil fuel industries and offering a more decentralized deployment model, may face fewer public resistance issues. However, transparency, stakeholder engagement, and clear communication will still be essential for building trust and ensuring that both technologies are viewed as credible and beneficial climate solutions.

Future Outlook and Strategic Synergies

The path forward involves leveraging the strengths of both DAC and point source capture. While point source technologies are critical for immediate emissions reductions, DAC provides a long-term mechanism for balancing hard-to-abate emissions and removing excess CO_2 from the atmosphere. In the coming decades, integrated carbon management systems that include both approaches will likely become standard. For example, an industrial complex may capture emissions at the source while also supporting a nearby DAC facility to offset residual emissions. Advancing both technologies in parallel, supported by strong policy frameworks and international cooperation, will be essential for achieving deep decarbonization and maintaining climate stability.

2.4.3 Commercialization of DAC Technologies

The increasing urgency to address global climate change has accelerated interest in carbon dioxide removal (CDR) technologies, particularly Direct Air Capture (DAC). Unlike conventional carbon capture methods that focus on point sources such as industrial facilities and power plants, DAC extracts carbon dioxide directly from the ambient atmosphere. This unique capability makes DAC a promising tool for achieving net-zero and even net-negative emissions. However, for DAC to significantly contribute to global climate mitigation goals, it must move beyond laboratory demonstrations and pilot projects into large-scale, economically viable deployment.

The commercialization of DAC technologies involves overcoming substantial technical, financial, infrastructural, and regulatory hurdles, while also capturing emerging market opportunities and ensuring public and environmental trust. The commercialization of Direct Air Capture represents a promising yet complex challenge in the fight against climate change. While technical breakthroughs have demonstrated the feasibility of extracting CO_2 directly from air, the journey to economic and operational viability is still unfolding. Key enablers include falling costs through innovation, policy incentives that create demand, robust infrastructure planning, and clear environmental standards. DAC offers a unique and essential function in climate mitigation by removing past and distributed emissions, but it cannot act alone. As part of a broader portfolio of decarbonization solutions, DAC has the potential to become a cornerstone of global efforts to reach net-zero and safeguard the planet's future.

Role of DAC in Climate Strategy

As climate science underscores the need for deep decarbonization, DAC has gained attention as a technology capable of removing excess atmospheric CO_2. Unlike emissions reduction strategies, DAC provides the means to actively reverse some of the accumulated emissions, making it essential for offsetting hard-to-abate sectors such as aviation, agriculture, and shipping. Leading climate models, including those by the Intergovernmental Panel on Climate Change (IPCC), consistently show that large-scale deployment of carbon removal technologies, including DAC, is necessary to meet the Paris Agreement's target of limiting global warming to 1.5 degrees Celsius. This scientific consensus has laid the foundation for increasing investment and interest in commercializing DAC.

Technology Landscape and Leading Companies

The current DAC landscape is dominated by a few pioneering companies that are advancing the technology through innovation and deployment. Climeworks, based in Switzerland, is one of the most prominent players and has developed modular DAC systems that use solid sorbents to capture CO_2. Another key company, Carbon Engineering from Canada, uses a liquid solvent approach and is working on scaling its technology in partnership with major energy firms. U.S.-based Global Thermostat is also developing commercial DAC systems using solid amine-based sorbents. These companies represent the leading edge of a growing ecosystem of start-ups, research institutions, and industrial partners aiming to bring DAC technologies to market.

Business Models and Revenue Streams

The commercialization of DAC requires viable business models that can support long-term operations and scaling. Currently, the most common revenue stream is the sale of carbon removal credits to corporations seeking to offset their emissions. These credits are part of voluntary carbon markets, where companies pay for the removal of a specific quantity of CO_2 to meet their sustainability goals. Some DAC firms also sell captured CO_2 for use in industries such as synthetic fuels, carbonated beverages, greenhouses, and construction materials. However, carbon utilization markets remain limited in scale and cannot absorb the gigaton-scale removals required for climate stabilization. As a result, long-term commercialization will rely on expanded carbon credit markets and potentially new government incentives or mandates for carbon removal.

Cost Considerations and Scaling Challenges

One of the biggest obstacles to commercializing DAC is the high cost of capturing CO_2 from ambient air. Current estimates range from \$600 to \$1,200 per ton of CO_2 captured, depending on the technology and scale of deployment. These costs are significantly higher than those associated with point source carbon capture, making DAC economically challenging without policy support. However, industry experts and researchers anticipate that costs will decline with scale, innovation, and learning-by-doing effects. For example, modular DAC units may benefit from mass manufacturing, while advances in sorbent materials and process efficiency can lower operational expenses. Still, reducing costs to below \$200 per ton—a threshold often cited as necessary for large-scale deployment—remains a key goal for the industry.

Infrastructure and Energy Requirements

DAC systems require significant infrastructure to operate at scale. They need vast arrays of fans, reactors, and regeneration systems, as well as energy inputs for sorbent heating and CO_2 compression. Because of these energy requirements, low-carbon or renewable energy sources are essential to ensure the net benefit of DAC technologies. Commercial deployment must also address CO_2 transportation and storage infrastructure, particularly for geological sequestration. Pipelines, compression stations, and injection wells will need to be developed or expanded in coordination with DAC facilities. Without a comprehensive carbon management infrastructure, the commercialization of DAC could face delays and increased costs.

Policy Support and Government Initiatives

Public policy plays a critical role in accelerating the commercialization of DAC. In the United States, the Inflation Reduction Act (IRA) significantly increased the value of the 45Q tax credit for carbon removal, offering up to \$180 per ton of CO_2 captured and permanently stored using DAC. This policy support has catalyzed new investments and project announcements. In Europe, the European Union's Innovation Fund and national programs have begun supporting DAC research and pilot projects. The United Kingdom and Canada are also exploring regulatory frameworks and funding mechanisms to promote DAC deployment. These policy tools are vital for closing the economic gap and making DAC competitive in carbon markets. Future policy innovations may include procurement mandates, inclusion in compliance carbon markets, and subsidies for low-carbon products made using captured CO_2.

Public Perception and Environmental Considerations

The success of DAC commercialization also depends on public perception and environmental acceptance. As a relatively new and high-tech solution, DAC is often met with skepticism or misconceptions. Critics argue that reliance on carbon removal could distract from the urgent need to reduce emissions. Others express concern about the energy footprint and resource use of large-scale DAC facilities. To gain public trust, companies and governments must prioritize transparency, stakeholder engagement, and rigorous environmental assessments. Demonstrating that DAC complements rather than replaces emissions reductions will be crucial for maintaining support. Moreover, integrating DAC into sustainable energy systems and minimizing environmental impacts will reinforce its legitimacy as a climate solution.

Market Demand and Corporate Climate Commitments

The growing number of corporations pledging to achieve net-zero emissions has created a burgeoning market for carbon removal services. Companies such as Microsoft, Shopify, Stripe, and Alphabet have already signed contracts to purchase DAC-based carbon removal credits, often at premium prices, to support early-stage deployment. These corporate commitments signal strong demand for high-quality carbon removal and provide critical early revenue for DAC developers. Over time, as the carbon credit market becomes more regulated and standardized, DAC projects that can offer verifiable, permanent removal will be well-positioned to lead. This corporate interest is not only helping commercialize DAC technologies but also demonstrating the feasibility of integrating carbon removal into private sector sustainability strategies.

Research and Development Pathways

Continued research and development will be essential to make DAC commercially viable at a large scale. Areas of focus include improving sorbent materials for higher capacity and durability, reducing energy requirements for regeneration, and optimizing system design for modularity and scalability. Collaboration between academic institutions, national laboratories, and private companies can accelerate these innovations. Governments can further support R&D through targeted grants, competitive funding calls, and public-private partnerships.

Advances in artificial intelligence, process engineering, and materials science also hold potential to improve DAC performance. As the field matures, standardizing measurement, verification, and reporting methodologies will ensure transparency and accountability in carbon removal claims.

Pathways to Commercial Maturity

The commercialization of DAC will likely follow a staged pathway, starting with small-scale pilot projects and expanding to regional and eventually global deployment. Early demonstration plants will prove technical feasibility and inform process improvements. Mid-scale facilities can benefit from economies of scale and attract commercial investment. Eventually, large-scale DAC hubs connected to low-carbon energy and CO_2 storage networks could become part of national carbon management strategies. Coordinated action across stakeholders—including technology developers, energy providers, regulators, financiers, and communities—will be necessary to enable this transition. Market readiness, cost reductions, and supportive policies will determine how quickly and successfully DAC can scale to the levels required for meaningful climate impact.

2.5 BIOENERGY WITH CARBON CAPTURE AND STORAGE (BECCS)

2.5.1 How BECCS Combines Biomass Energy and CCS

Bioenergy with Carbon Capture and Storage, commonly known as BECCS, is an innovative approach that integrates renewable energy generation from biomass with carbon capture and storage technologies. This method has gained significant attention in recent years as a potentially vital solution in achieving net-negative greenhouse gas emissions. As the world faces the urgent challenge of reducing atmospheric carbon dioxide to meet climate targets set by the Paris Agreement, BECCS offers a promising pathway. By generating energy through the combustion or processing of biological materials and capturing the carbon dioxide emissions produced, BECCS helps to remove carbon from the atmosphere, storing it in secure geological formations.

This technology represents a unique blend of renewable energy and carbon management, capable of contributing to long-term climate mitigation strategies. BECCS represents a powerful convergence of bioenergy and carbon capture technologies with the potential to deliver net-negative emissions and support global climate mitigation efforts. By utilizing biomass as a renewable energy source and capturing the resulting CO_2 emissions for permanent storage, BECCS can effectively remove carbon from the atmosphere while supplying energy. Although challenges related to cost, sustainability, and deployment remain, growing recognition of BECCS's role in achieving climate targets is driving interest and investment. With the right combination of innovation, policy support, and public engagement, BECCS can play a transformative role in shaping a low-carbon, sustainable future.

Principles of BECCS

The concept behind BECCS begins with the use of biomass as an energy source. Biomass refers to organic material derived from plants, trees, agricultural residues, or dedicated energy crops. As these plants grow, they absorb carbon dioxide from the atmosphere through the process of photosynthesis. When the biomass is used to produce energy—typically through combustion, gasification, or fermentation—carbon dioxide is released. In a traditional bioenergy system, this CO_2 would be emitted back into the atmosphere, creating a closed-loop cycle with net-zero emissions. However, BECCS enhances this process by capturing the released CO_2 before it escapes into the air and transporting it for permanent storage underground. This intervention turns a neutral cycle into a carbon-negative one, effectively removing greenhouse gases from the atmosphere.

Biomass Sources and Energy Conversion Methods

BECCS relies on a variety of biomass sources, ranging from forest residues and agricultural waste to purpose-grown energy crops such as switchgrass, willow, and miscanthus. Additionally, industrial and municipal organic waste can also be utilized, contributing to waste management alongside energy production. The method of energy conversion plays a key role in determining the type and volume of CO_2 emissions that can be captured. Combustion of biomass in power plants generates heat and electricity, while fermentation processes, such as those used in ethanol production, produce biofuels. Each of these processes offers different points for capturing CO_2. For example, in fermentation, CO_2 is released in a relatively pure form, making it easier to capture and compress. In combustion or gasification, post-combustion or pre-combustion capture techniques are used to isolate the carbon emissions from flue gases or synthesis gas.

Carbon Capture Techniques in BECCS

The carbon capture component of BECCS typically employs one of three approaches: post-combustion, pre-combustion, or oxy-fuel combustion. In post-combustion capture, CO_2 is extracted from flue gases after biomass combustion using solvents, membranes, or solid sorbents. Pre-combustion capture involves gasifying the biomass to produce syngas, which is then processed to separate the CO_2 before the fuel is combusted. Oxy-fuel combustion, on the other hand, burns the biomass in pure oxygen, producing a concentrated stream of CO_2 that simplifies capture.

These technologies have been developed in the context of fossil fuel-based carbon capture and are now being adapted for bioenergy applications. Once captured, the CO_2 is compressed and transported, typically via pipelines, to storage sites where it can be injected deep underground into geological formations such as saline aquifers, depleted oil and gas fields, or basalt rock reservoirs.

Carbon Storage and Permanence

The final step in the BECCS process is the long-term storage of the captured carbon dioxide. Geological storage is the most widely adopted method and involves injecting CO_2 into deep underground formations where it is trapped by layers of impermeable rock. Over time, the CO_2 becomes mineralized or dissolved in subsurface fluids, ensuring its stability. The success of BECCS in delivering negative emissions depends largely on the permanence and security of this storage. Monitoring systems are essential to detect potential leaks and ensure compliance with environmental regulations. Existing carbon storage projects around the world have demonstrated that with the proper geological conditions and management practices, CO_2 can be stored safely and effectively for thousands of years.

BECCS and Net-Negative Emissions

One of the most compelling features of BECCS is its ability to produce net-negative emissions. Unlike conventional mitigation strategies that reduce the rate of CO_2 emissions, BECCS actively removes CO_2 from the atmosphere and stores it. This makes BECCS a key candidate in climate models that aim to limit global warming to 1.5 or 2 degrees Celsius above pre-industrial levels. The Intergovernmental Panel on Climate Change (IPCC) and other leading organizations have highlighted BECCS as a necessary component in scenarios that require carbon dioxide removal at the gigaton scale. By providing a renewable source of energy while simultaneously reducing atmospheric CO_2, BECCS offers dual benefits that align with sustainable development goals.

Current Deployment and Pilot Projects

Although the theoretical potential of BECCS is considerable, actual deployment remains limited. As of now, only a handful of BECCS facilities are operational globally, most of which are small-scale or demonstration projects. The most notable example is the Archer Daniels Midland (ADM) facility in Illinois, USA, which captures CO_2 from bioethanol production and stores it in a saline aquifer. Other projects are in various stages of planning or construction in countries such as the United Kingdom, Sweden, and Japan. The low number of commercial BECCS projects is due to several barriers, including high capital costs, technological complexity, and a lack of clear policy incentives for negative emissions.

Economic and Policy Considerations

The commercialization of BECCS faces significant economic challenges. High upfront investment, operational costs for capture and storage, and uncertain revenue streams make it difficult for BECCS projects to compete with other forms of energy. Government intervention in the form of subsidies, tax credits, carbon pricing, and regulatory support is essential to encourage the adoption of BECCS. Policies such as the U.S. 45Q tax credit, the EU Innovation Fund, and the UK's Net Zero Strategy are beginning to create more favorable conditions for BECCS. However, further policy clarity and international cooperation will be required to stimulate large-scale deployment and attract private sector investment.

Environmental and Sustainability Implications

The large-scale use of biomass for energy production raises important environmental and sustainability concerns. The cultivation of energy crops may lead to land-use change, deforestation, and competition with food production, potentially undermining the carbon benefits of BECCS. To avoid these negative outcomes, it is crucial to adopt sustainable biomass sourcing practices, such as using agricultural residues, marginal lands, or integrating agroforestry systems. Water use, biodiversity impacts, and the carbon footprint of transporting biomass also need to be considered in evaluating the overall sustainability of BECCS. Lifecycle assessments can help determine the net climate benefit and guide responsible deployment.

Public Perception and Social Acceptance

Social acceptance is a critical factor in the successful implementation of BECCS projects. Public concerns about land use, biomass harvesting, and underground CO_2 storage can hinder project approval and development. Transparent communication, community engagement, and inclusive decision-making processes are necessary to build trust and address potential risks. Highlighting the climate benefits of BECCS and its role in achieving national and global climate targets can help gain public support. Demonstrating successful case studies and ensuring equitable access to the benefits of BECCS, such as job creation and rural development, will also enhance social acceptance.

2.5.2 Potential for Negative Emissions

As the global community intensifies efforts to limit global temperature rise in accordance with the Paris Agreement, the concept of negative emissions has taken center stage in climate policy discussions. Negative emissions refer to the removal of carbon dioxide from the atmosphere and its long-term sequestration, thereby reducing the overall concentration of greenhouse gases. Among the technologies capable of delivering such outcomes, Bioenergy with Carbon Capture and Storage

(BECCS) holds a unique and promising position. BECCS combines the generation of energy from biomass with the capture and storage of carbon dioxide that is released during energy production. This integration creates a closed-loop system where CO_2 previously absorbed from the atmosphere by plants is permanently stored underground, resulting in a net reduction of atmospheric carbon levels.

Mechanism of Negative Emissions in BECCS

The foundation of BECCS lies in the natural process of photosynthesis, through which plants absorb carbon dioxide from the air and convert it into biomass. When this biomass is used as fuel in energy production, it releases the captured CO_2 back into the atmosphere. However, with BECCS, this CO_2 is captured before it escapes into the air and is then transported to geological storage sites where it is permanently sequestered. This transforms what would otherwise be a carbon-neutral cycle into one that is carbon-negative. Unlike conventional carbon capture and storage (CCS) applied to fossil fuels, BECCS leverages the biological sequestration capabilities of biomass, making it a renewable and sustainable option for removing CO_2 from the atmosphere.

Theoretical Potential of BECCS for Climate Mitigation

Climate models, especially those developed by the Intergovernmental Panel on Climate Change (IPCC), have identified BECCS as one of the most scalable technologies for achieving negative emissions. According to several integrated assessment models, BECCS could contribute between 3 to 11 gigatonnes of CO_2 removal per year by the end of the 21st century, depending on deployment rates and resource availability. This capacity is critical for offsetting emissions from hard-to-abate sectors such as aviation, shipping, and heavy industry. Furthermore, BECCS offers the advantage of providing carbon-negative power, heat, or biofuels, thereby contributing simultaneously to decarbonizing the energy sector and achieving net-negative emissions.

Current Implementation and Technological Maturity

While the theoretical promise of BECCS is well-documented, real-world implementation remains limited. Only a few operational projects exist globally, such as the Archer Daniels Midland facility in Illinois, which captures CO_2 from bioethanol production and stores it in a saline aquifer. Most existing applications are pilot-scale or focused on specific sectors like biofuels rather than large-scale power generation. Technological components of BECCS, including biomass combustion, fermentation, gasification, and CO_2 capture technologies, are individually mature, but their integration into a cohesive and efficient system at scale requires further development. Continued investment in research, infrastructure, and demonstration projects is essential to accelerate the deployment of BECCS technologies.

Sustainability of Biomass Feedstocks

The potential for negative emissions through BECCS depends heavily on the sustainability of biomass sourcing. Unsustainable practices such as deforestation or land-use change for energy crop cultivation can negate the carbon removal benefits of BECCS and may even result in net-positive emissions. To maintain the integrity of BECCS as a negative emissions technology, biomass must be sourced responsibly. This involves utilizing agricultural residues, forestry by-products, municipal organic waste, or cultivating energy crops on degraded or marginal lands that do not compete with food production. Sustainability certification schemes, life cycle assessments, and land-use planning are critical tools to ensure that biomass supply chains contribute positively to climate goals.

Geological Storage and Permanence of Sequestration

Another determinant of BECCS's effectiveness is the permanence and security of carbon storage. The captured CO_2 must be injected into deep geological formations where it can remain trapped for thousands of years. Suitable storage sites include depleted oil and gas reservoirs, deep saline aquifers, and basalt formations. Monitoring and verification technologies are employed to ensure that stored CO_2 does not leak and that environmental safety is maintained. Countries like Norway, the United States, and Canada have successfully demonstrated the feasibility of long-term CO_2 storage through existing CCS projects. The infrastructure and regulatory frameworks established in these regions can serve as models for future BECCS deployment.

Economic Considerations and Policy Support

The economic viability of BECCS is currently one of its major challenges. The cost of capturing, transporting, and storing CO_2, combined with the logistics of biomass supply, results in higher energy prices compared to conventional power generation. Without adequate policy support or market incentives, BECCS is unlikely to achieve widespread adoption. Mechanisms such as carbon pricing, tax credits, subsidies, and negative emissions trading schemes can help internalize the climate benefits of BECCS and improve its competitiveness. Policies like the U.S. 45Q tax credit and the European Union's Emissions Trading System are early examples of initiatives that could facilitate the commercialization of BECCS and other negative emissions technologies.

Comparison with Other Negative Emissions Technologies

While BECCS is one of the leading options for carbon dioxide removal, it is not the only solution. Other negative emissions technologies include Direct Air Capture (DAC), afforestation and reforestation, enhanced weathering, and soil carbon sequestration. Each of these options varies in terms of scalability, cost, land use, and permanence. For instance, afforestation is low-cost but limited by land availability and risks such as forest fires. DAC offers higher permanence but is currently energy-intensive and expensive. BECCS occupies a unique niche by providing both energy and carbon removal but requires careful management to avoid adverse land and ecosystem impacts. A diversified portfolio of negative emissions solutions will likely be necessary to achieve global climate targets.

Environmental and Social Implications

In addition to carbon removal, BECCS has broader environmental and social implications. On the positive side, it can contribute to rural development, energy access, and job creation in biomass-producing regions. However, large-scale biomass production can pose risks to biodiversity, water resources, and food security if not managed properly. Moreover, the social acceptance of BECCS projects depends on transparent communication, stakeholder engagement, and equitable distribution of benefits and risks. Integrating local communities into planning and decision-making processes is essential to build trust and ensure that BECCS contributes to sustainable development.

Integration with National and Global Climate Strategies

BECCS has been incorporated into many national climate plans and long-term strategies as a tool for balancing residual emissions and achieving net-zero targets. Countries such as the United Kingdom, Sweden, and Japan have identified BECCS as a strategic component of their decarbonization pathways. At the global level, scenarios consistent with the 1.5°C target often rely on significant BECCS deployment, especially in the second half of the century. The alignment of BECCS development with national policy frameworks, investment plans, and international cooperation is necessary to ensure its scalability and effectiveness. Multilateral funding mechanisms and technology transfer initiatives can also play a vital role in supporting BECCS implementation in developing countries.

2.5.3 BECCS Deployment Challenges

While BECCS offers a powerful tool for negative emissions and a potential pathway to achieving net-zero and even net-negative climate targets, its practical deployment is fraught with significant challenges. From biomass sustainability and technological integration to infrastructure, policy, and social acceptance, the barriers to BECCS implementation are complex and interrelated. Overcoming these obstacles will require coordinated action across governments, industries, and civil society, supported by rigorous research, robust policies, and sustained investment. Ultimately, BECCS can play a vital role in global climate mitigation efforts, but only if its deployment is guided by careful planning, equitable governance, and a commitment to sustainability.

Biomass Supply and Sustainability Concerns

One of the foremost challenges in deploying BECCS is the sourcing and sustainability of biomass feedstocks. To achieve large-scale deployment, enormous quantities of biomass are needed, which raises significant questions about land use, water consumption, biodiversity impacts, and competition with food production. Unsustainable practices such as deforestation or monoculture energy crop plantations could result in carbon emissions that negate the benefits of BECCS. Additionally, transporting large volumes of biomass over long distances can further increase the carbon footprint of the system. Ensuring the sustainability of biomass supply chains is therefore critical, requiring stringent criteria, certifications, and regulations to ensure that BECCS does not compromise ecological integrity or food security.

High Costs and Economic Viability

The economic challenge of BECCS deployment is substantial. Capturing carbon dioxide, transporting it, and storing it underground are capital-intensive processes that increase the overall cost of energy production. Unlike fossil-based CCS, which can sometimes be offset by enhanced oil recovery revenues, BECCS does not generally generate direct financial returns from the storage of carbon. Without robust carbon pricing mechanisms or government subsidies, BECCS projects struggle to compete with other low-carbon energy sources such as wind or solar. Moreover, establishing the necessary infrastructure for CO_2 transport and storage, including pipelines and storage sites, requires significant upfront investment and long-term financial commitment.

Technological Integration and Maturity

Although the individual components of BECCS—biomass energy production, carbon capture, and geological storage—are commercially available technologies, their integrated deployment at scale presents technical difficulties. Variations in biomass type, combustion or gasification processes, and carbon capture systems create operational complexity.

Scaling up BECCS requires optimized and flexible systems that can handle diverse feedstocks and deliver reliable performance over time. Furthermore, many existing biomass plants are not designed with carbon capture in mind and would require expensive retrofitting. Continued research and development are necessary to improve the efficiency, cost-effectiveness, and operational integration of BECCS technologies.

Infrastructure and Storage Capacity Limitations

Effective deployment of BECCS depends heavily on the availability of infrastructure for carbon transport and storage. Suitable geological storage sites must be identified, assessed for safety and capacity, and developed with robust monitoring and verification systems. In many regions of the world, such infrastructure does not yet exist, and building it would require considerable investment and planning. Moreover, regulatory frameworks for carbon storage are still evolving, and legal uncertainties about long-term liability for stored CO_2 may deter private investment. The co-location of biomass sources, power plants, and storage sites is also rare, often necessitating extensive transportation networks that further increase costs and complexity.

Policy Gaps and Regulatory Uncertainty

Policy and regulatory uncertainty remain major barriers to BECCS deployment. Although BECCS features prominently in international climate models and scenarios, there is a lack of concrete national-level strategies, incentives, and regulations to support its development. In many jurisdictions, policies either do not recognize negative emissions or fail to distinguish between different carbon removal methods. This creates an uneven playing field and inhibits investment. Clear policy direction, supportive regulatory environments, and stable long-term incentives such as carbon pricing, tax credits, and funding for demonstration projects are essential to move BECCS from pilot scale to widespread deployment. Governments must also establish legal frameworks to govern CO_2 transport, storage rights, and long-term monitoring responsibilities.

Land Use and Social Acceptance Issues

The land use implications of BECCS represent another significant challenge. Deploying BECCS at scale requires vast areas of land for biomass cultivation, which can compete with food production, lead to displacement of communities, and affect land tenure systems. These impacts raise serious ethical and equity concerns, especially in developing countries where land is already a contested resource. Additionally, the infrastructure associated with BECCS, such as pipelines and storage sites, may face opposition from local communities due to perceived environmental and safety risks. Gaining public acceptance for BECCS projects requires transparent communication, community engagement, and equitable benefit-sharing mechanisms. Social license to operate will be crucial for long-term success.

Environmental Risks and Trade-Offs

Although BECCS is intended to reduce atmospheric carbon dioxide levels, it is not without environmental risks. Large-scale biomass cultivation can lead to deforestation, habitat loss, and soil degradation. The use of fertilizers and water for energy crops may contribute to water pollution and resource depletion. In addition, the carbon storage process itself must be carefully managed to prevent leakage and ensure permanence. If CO_2 were to leak from geological storage sites, it could pose risks to ecosystems and human health and undermine the climate benefits of BECCS. These trade-offs necessitate a thorough life cycle assessment of BECCS systems and careful site selection and monitoring to mitigate environmental impacts.

Competition with Other Carbon Removal Approaches

BECCS does not operate in isolation but is one among several negative emissions technologies being considered to achieve climate goals. Direct Air Capture (DAC), afforestation, enhanced weathering, and soil carbon sequestration are all alternative or complementary approaches to carbon removal. In terms of cost, scalability, and environmental impact, each of these methods has distinct advantages and limitations. As such, BECCS may face competition for policy support and funding. Policymakers will need to carefully assess the relative merits of different options and design integrated strategies that balance multiple pathways to carbon neutrality, rather than relying exclusively on any single technology.

Uncertainty in Climate Modeling and Over-Reliance on BECCS

Another important challenge is the potential over-reliance on BECCS in climate models and policy roadmaps. Many scenarios assume large-scale deployment of BECCS in the second half of the century to offset residual emissions, but these projections may not adequately account for the challenges described above. If BECCS fails to scale as expected, it could jeopardize the achievement of climate targets and create a false sense of security. Climate strategies must therefore remain flexible and diversified, investing in mitigation options that are already feasible while continuing to develop BECCS and other negative emissions technologies. Realistic expectations and contingency planning are essential to avoid policy failures.

Capacity Building and International Collaboration

The deployment of BECCS at the scale necessary for meaningful climate impact will require international collaboration and capacity building, particularly in developing countries. Many of these countries possess abundant biomass resources and storage potential but lack the technical expertise, financial resources, and institutional capacity to implement BECCS. Technology transfer, financial assistance, and collaborative research are crucial for global equity and effectiveness. International climate finance mechanisms such as the Green Climate Fund can play a role in supporting BECCS development in low-income regions, ensuring that benefits are widely shared and that no country is left behind in the transition to a low-carbon future.

CHAPTER

Carbon Transport Systems

3

3.1 OVERVIEW OF CO_2 TRANSPORT METHODS

3.1.1 Pipeline Transport of CO_2

Carbon dioxide (CO_2) pipeline transport plays a critical role in the chain of Carbon Capture and Storage (CCS), enabling the transfer of captured CO_2 from industrial sources to designated storage sites. As CCS systems are increasingly seen as a necessary tool for mitigating climate change, the importance of reliable, safe, and efficient transport infrastructure has come into sharper focus. Among the available transport options, pipelines offer a cost-effective and scalable solution for moving large volumes of CO_2 over long distances. The development, deployment, and regulation of CO_2 pipelines are vital to the broader success of CCS technologies and achieving long-term climate goals. Pipeline transport of CO_2 is an essential link in the CCS chain, enabling captured carbon to be moved from emission sources to secure storage locations efficiently and safely. Drawing on decades of experience, the technology for CO_2 pipelines is mature and well-understood, but scaling it up for climate mitigation requires coordinated policy, investment, and stakeholder engagement. By addressing the technical, economic, regulatory, and social challenges associated with CO_2 transport, the world can unlock the full potential of CCS and move closer to a sustainable, low-carbon future. Pipeline infrastructure, if planned and implemented wisely, will serve as the arterial network through which carbon-neutral and eventually carbon-negative ambitions are realized.

Historical Context and Development

The concept of transporting CO_2 via pipelines is not new. The first CO_2 pipelines were constructed in the United States in the 1970s, primarily for use in enhanced oil recovery (EOR) operations. These pipelines transported CO_2 to oil fields where it was injected underground to boost oil production. Over time, this infrastructure laid the groundwork for the use of CO_2 pipelines in CCS systems. With the increasing urgency to reduce greenhouse gas emissions, the focus has shifted from using CO_2 for commercial purposes to ensuring its permanent and safe storage underground. The accumulated experience from decades of EOR has helped inform the design, safety standards, and operational procedures of pipelines used in modern CCS systems.

Design and Engineering of CO_2 Pipelines

CO_2 pipelines are typically designed to carry CO_2 in its dense or supercritical phase, where it behaves like both a liquid and a gas. This state allows for efficient transportation under high pressure and relatively low temperature. Pipeline design must account for several factors, including pressure management, corrosion resistance, temperature control, and the specific characteristics of the CO_2 stream, which may contain impurities like water, hydrogen sulfide, or nitrogen. Materials used in the construction of CO_2 pipelines are carefully selected to withstand corrosion and maintain structural integrity over extended periods. Engineers also design systems to monitor flow rates, pressure levels, and possible leaks, ensuring that safety and efficiency are maintained throughout the transport process.

Route Planning and Infrastructure Development

Planning the route for a CO_2 pipeline involves multiple considerations including geography, land use, environmental impacts, and existing infrastructure. Developers must identify paths that minimize disruptions to human settlements, ecosystems, and economic activities. In many cases, pipelines follow existing corridors used by oil, gas, or water pipelines to reduce environmental and social disruption. However, securing land rights, negotiating with landowners, and navigating complex regulatory frameworks are often significant hurdles. Infrastructure development also involves collaboration with public authorities, environmental agencies, and local communities to ensure that concerns are addressed and that the project aligns with public interests.

Safety Considerations and Risk Management

Safety is a central concern in CO_2 pipeline operations. While CO_2 is non-flammable and non-toxic in low concentrations, it can pose health risks in high concentrations and may displace oxygen in confined areas. A pipeline rupture or leak could result in a rapid release of pressurized gas, potentially leading to asphyxiation hazards, environmental damage, and public health concerns. To mitigate these risks, pipelines are equipped with advanced monitoring systems that detect changes in pressure, flow, and temperature, triggering automatic shutdowns or alerts in case of anomalies. Regular maintenance, rigorous inspection protocols, and adherence to international safety standards further reduce the likelihood of incidents. In addition, emergency response plans are established to manage potential leaks or accidents swiftly and effectively.

Economic and Logistical Advantages

Pipelines provide a highly economical mode of transporting CO_2, particularly when large volumes are involved or when transport distances are long. Compared to road or rail transport, pipelines offer lower operational costs and higher efficiency once the infrastructure is in place. They also reduce greenhouse gas emissions associated with alternative transport modes, thus enhancing the overall environmental benefits of CCS. Pipelines can operate continuously, providing a stable and reliable transport mechanism that matches the steady output of large industrial CO_2 sources such as power plants, cement factories, and chemical processing units. Over time, as more CCS projects are implemented, economies of scale are expected to lower the unit cost of pipeline transport even further.

Regulatory Framework and Policy Support

The development and operation of CO_2 pipelines are governed by national and regional regulations that ensure public safety, environmental protection, and economic fairness. In jurisdictions like the United States, agencies such as the Pipeline and Hazardous Materials Safety Administration (PHMSA) regulate the construction, maintenance, and safety standards of CO_2 pipelines. Other countries are developing or revising their regulatory frameworks to accommodate the growing role of CCS. Clear and harmonized regulations help reduce uncertainty for investors and project developers, enabling more predictable and efficient project implementation. Additionally, supportive policies such as subsidies, tax incentives, and carbon pricing mechanisms can improve the economic feasibility of CO_2 pipeline networks and accelerate their deployment.

Integration into CCS Networks

As CCS scales up globally, CO_2 pipelines are increasingly viewed not just as isolated infrastructure elements but as part of broader integrated networks. These networks could include multiple capture sources, shared transport routes, and centralized storage hubs. Such integration would increase efficiency, reduce redundancy, and lower overall costs. The development of CO_2 pipeline networks also allows for future expansion and interconnection between regions, creating a flexible system that can adapt to changing emission patterns and storage availability. In this vision, CO_2 pipelines function much like today's natural gas networks, providing a backbone for the decarbonized industrial economy of the future.

Challenges and Barriers to Expansion

Despite their advantages, CO_2 pipelines face several challenges that must be addressed for widespread adoption. High upfront capital costs, complex permitting processes, and public resistance can delay or derail projects. Building new pipelines often requires navigating a maze of regulatory approvals, environmental assessments, and legal challenges. In some regions, insufficient geological storage data or lack of established storage sites further complicates the planning of pipeline routes. Public opposition, fueled by fears of safety risks or environmental harm, may also arise, especially if communities feel inadequately informed or compensated. Overcoming these barriers requires proactive stakeholder engagement, transparent communication, and fair compensation practices.

3.1.2 CO_2 Shipping and Maritime Transport

Carbon dioxide (CO_2) shipping and maritime transport is becoming an increasingly important component of global Carbon Capture and Storage (CCS) strategies. While pipelines are the most commonly used method for transporting large volumes of CO_2, shipping offers flexibility, scalability, and international reach, making it particularly valuable in certain geographical and logistical contexts. CO_2 shipping enables the movement of captured carbon from industrial sources to storage locations that may be far away or located in other countries. As interest in decarbonization grows, especially across industries that emit large volumes of CO_2, maritime transport is gaining recognition as a critical infrastructure solution for achieving net-zero goals.

CO_2 shipping and maritime transport is an emerging but essential component of the carbon management ecosystem. It offers a practical and scalable solution for connecting carbon capture facilities to storage sites across distances and national boundaries. With the right investments in technology, infrastructure, regulation, and stakeholder coordination, shipping can become a backbone of global CCS networks. By leveraging maritime transport, the world can accelerate the deployment of CCS, support hard-to-abate sectors, and move closer to achieving net-zero emissions. As climate action becomes increasingly urgent, CO_2 shipping stands out as a critical enabler of the low-carbon transition.

CO_2 Shipping

The basic idea behind CO_2 shipping is relatively straightforward. Captured carbon dioxide is liquefied and loaded into specially designed tanks aboard ships. These vessels then transport the liquefied CO_2 to storage sites, typically located near depleted oil and gas fields or in deep saline aquifers. The liquid form is favored because it allows a higher volume of CO_2 to be stored in a relatively small space, making maritime transport both practical and efficient. Shipping is particularly useful in areas where pipeline construction is difficult, cost-prohibitive, or not yet justified due to the scale of captured CO_2.

Advantages of Maritime CO_2 Transport

Maritime CO_2 transport offers a number of advantages that make it an attractive option in the broader CCS value chain. One of its most significant strengths is flexibility. Ships can be rerouted, adapted for multiple destinations, and used for variable volumes, which is not possible with fixed pipeline routes. This makes shipping ideal for early-stage CCS projects that have not yet scaled up to justify pipeline investment. Furthermore, ships can connect CO_2 sources in landlocked or remote areas to offshore storage sites. In this way, shipping supports international collaboration in carbon management and facilitates the development of regional

CO_2 hubs where capture, transport, and storage infrastructure can be shared across national boundaries.

Technological Aspects of CO_2 Shipping

Transporting CO_2 by sea involves several specialized technologies. Before loading, CO_2 is compressed and cooled to transform it into a liquid state, typically at a pressure of 6 to 7 bar and a temperature of around -50°C. It is then stored in cryogenic tanks made from materials that can withstand these extreme conditions. These tanks are often insulated and designed to minimize evaporation or leakage. During transport, temperature and pressure must be constantly monitored and maintained to ensure the liquid CO_2 remains stable. Unloading at the storage site also requires specific infrastructure, including pumps and pipelines that can handle liquefied gas safely and efficiently. Safety is a primary consideration at all stages, and the design of CO_2 ships follows strict international standards for cryogenic gas carriers.

Existing Experience and Pilot Projects

Although CO_2 shipping on a commercial scale is still emerging, there is relevant experience from industries that transport liquefied gases, such as liquefied natural gas (LNG) and liquefied petroleum gas (LPG). These industries have developed the maritime technologies, regulations, and best practices that can be adapted for CO_2. Some pilot projects and demonstration initiatives are already underway to explore the feasibility and optimize the logistics of CO_2 shipping. For instance, Northern Lights in Norway is among the first full-scale projects planning to transport CO_2 by ship from multiple capture sites in Europe to an offshore storage facility in the North Sea. Such projects are critical in establishing proof-of-concept and creating a knowledge base for wider deployment.

Economic Considerations and Cost Implications

The economics of CO_2 shipping depend on a range of factors, including transport distance, shipment volume, port infrastructure, and ship size. While pipelines are generally more cost-effective for continuous and high-volume transport over land, shipping becomes competitive or even preferable for long distances, small-to-medium volumes, or transport across water bodies.

In early phases of CCS deployment, shipping allows projects to proceed without the massive upfront investments required for pipelines. Additionally, shared shipping routes can lower costs through economies of scale. As technology improves and the scale of operations grows, it is expected that CO_2 shipping will become even more cost-efficient, contributing positively to the overall financial viability of CCS initiatives.

Regulatory and Legal Framework

For maritime CO_2 transport to scale, an appropriate regulatory and legal framework is essential. The international nature of shipping adds layers of complexity, as CO_2 may be transported between different jurisdictions with differing environmental, safety, and liability standards. International bodies such as the International Maritime Organization (IMO) are working to address these issues, especially with respect to safety regulations, environmental protection, and ship classification. Additionally, CO_2 is being considered for classification as a non-toxic, non-flammable cargo under international shipping codes. National laws must also be updated to facilitate permitting, customs, and cross-border CO_2 transfers. Clear legal guidelines will be necessary to reduce uncertainty for investors and operators alike.

Environmental and Safety Concerns

Although CO_2 shipping is seen as environmentally beneficial due to its role in carbon sequestration, there are still environmental and safety concerns that must be managed. Leaks during transport could result in localized CO_2 buildup, potentially posing health risks or contributing temporarily to atmospheric carbon. Therefore, robust monitoring and safety systems are mandatory. These include leak detection sensors, emergency shut-off systems, and double-walled containment structures. Environmental impact assessments must also consider emissions from the ships themselves, though the carbon intensity of maritime CO_2 transport is generally low relative to its climate benefits. Looking forward, the adoption of low-emission or zero-emission vessels will further enhance the sustainability of CO_2 shipping operations.

Infrastructure Requirements and Port Readiness

Ports play a central role in maritime CO_2 transport. For shipping to function efficiently, ports must be equipped with specialized infrastructure to handle liquefied CO_2. This includes loading and unloading systems, cryogenic storage tanks, safety systems, and dedicated berths. Port operators and local authorities must coordinate with CCS developers to ensure the timely development of such infrastructure. Port readiness will be a key determinant in the success of early projects and the pace at which shipping networks can expand. Investments in port infrastructure will need to be synchronized with shipbuilding programs and the development of capture and storage facilities, requiring comprehensive planning and cross-sector cooperation.

Integration with Global CCS Networks

CO_2 shipping has the potential to play a major role in the development of international CCS networks. These networks aim to link multiple sources of CO_2 emissions with suitable geological storage sites, often across national borders. Shipping provides the mobility and flexibility necessary to create such integrated systems, especially in regions lacking suitable storage geology or land-based transport routes. For example, small countries with heavy industries but no domestic storage potential can ship CO_2 to larger nations with ample offshore storage capacity. In this way, maritime CO_2 transport can support equitable and collaborative approaches to climate mitigation, enabling all countries to participate in global decarbonization efforts.

3.1.3 Compressed CO_2 and Liquefaction Technologies

The increasing urgency of climate change has placed a spotlight on carbon capture, utilization, and storage (CCUS) as a critical technology pathway for reducing atmospheric CO_2 emissions. One of the key technical stages in this process involves compressing and liquefying captured carbon dioxide to facilitate efficient transportation and long-term storage. The technologies used in compressing and liquefying CO_2 are central to the overall viability of carbon management, as they directly influence energy requirements, infrastructure design, and operational costs. Compressed CO_2 and liquefaction technologies are indispensable components of modern carbon capture and storage strategies.

By enabling the safe, efficient, and scalable transport of carbon dioxide from emission sources to storage sites, these systems bridge a critical gap in the carbon management value chain. While energy demands and technical challenges remain, continued innovation and integration with renewable energy sources promise to enhance the effectiveness and sustainability of these technologies. As the world moves toward deep decarbonization, mastering the science and engineering of CO_2 compression and liquefaction will be essential for achieving global climate objectives.

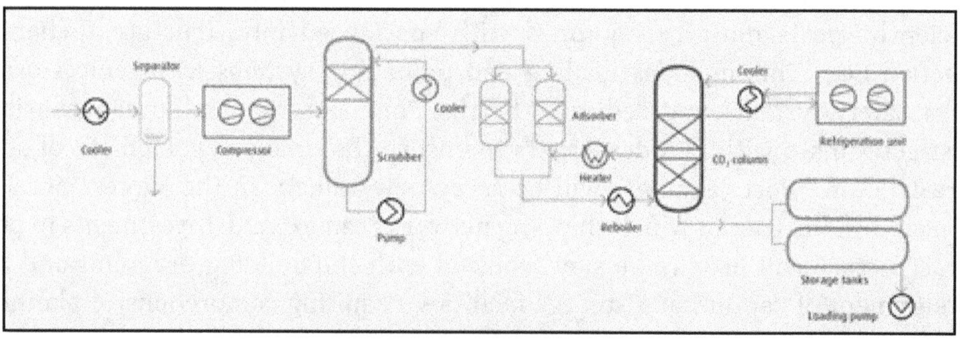

Fig: CO_2 Liquefaction Plant

Fundamentals of CO_2 Compression

After carbon dioxide is captured from industrial processes or power plants, it must be compressed from atmospheric pressure to a much higher pressure suitable for transport and storage. This compression reduces the volume of the gas significantly, allowing it to be moved more efficiently through pipelines or loaded into containers for shipping. The compression process typically brings the CO_2 to a pressure of around 100 to 150 bar for pipeline transport. The process occurs in multiple stages to prevent overheating and to optimize energy usage. Between each stage, intercoolers are used to remove heat generated during compression, improving efficiency and protecting mechanical components from thermal damage. CO_2 behaves differently from air or other gases during compression due to its unique phase behavior. Near its critical point, which is 31.1°C and 73.8 bar, CO_2 can exist as a supercritical fluid with properties of both a gas and a liquid. This supercritical state is highly desirable for storage and transport, as it combines the high density of a liquid with the low viscosity of a gas, facilitating ease of movement through pipelines. However, reaching and maintaining supercritical conditions requires precise control of temperature and pressure, demanding advanced engineering and robust monitoring systems.

Liquefaction of Carbon Dioxide

Liquefaction is another key process that transforms CO_2 into a dense, liquid state for storage or maritime transportation. Unlike compression for pipeline use, liquefaction targets even higher densities by cooling CO_2 to temperatures below -50°C and compressing it to a moderate pressure of about 6 to 7 bar. This combination of low temperature and pressure allows CO_2 to remain stable in its liquid form. Liquefaction is particularly important for ship-based transport and storage in cryogenic tanks, as it maximizes the volume of CO_2 that can be carried and stored within a given space. The technology for CO_2 liquefaction is similar in principle to that used for liquefied natural gas (LNG). However, it must be tailored to the specific thermodynamic characteristics of CO_2. The process typically involves refrigeration cycles using ammonia or propane as refrigerants. Multi-stage cooling systems progressively reduce the temperature of the CO_2, condensing it into liquid without freezing. Maintaining constant and efficient cooling is crucial, as temperature fluctuations can lead to partial vaporization, creating pressure imbalances or reducing storage capacity. The liquefied CO_2 must also be insulated during storage and transport to prevent thermal ingress and subsequent boiling.

Technological Components and Engineering Design

Both compression and liquefaction systems depend on a range of specialized equipment. For compression, multistage centrifugal or reciprocating compressors are commonly used, chosen based on flow rate, operating conditions, and energy efficiency. Intercoolers, aftercoolers, and condensers are integrated into the compression chain to manage heat and moisture content. Dry CO_2 is preferable to avoid corrosion and hydrate formation, so dehydration units are also employed to remove water vapor. For liquefaction, refrigeration systems include compressors, evaporators, heat exchangers, and control systems designed to handle cryogenic operations. Storage tanks for liquefied CO_2 are typically double-walled and vacuum-insulated, designed to handle thermal stress and prevent leakage. The entire system must be sealed and equipped with safety mechanisms such as pressure relief valves, leak detection sensors, and emergency shutdown protocols. Due to the high energy intensity and operational complexity of these systems, automation and process control technologies play a crucial role in maintaining stable and efficient operations.

Energy Requirements and Efficiency Challenges

One of the primary challenges associated with CO_2 compression and liquefaction is the high energy demand. Compressing CO_2 to supercritical or liquefied states consumes a significant portion of the total energy used in the CCS chain. For example, compression alone can account for up to 15 percent of the total energy output of a power plant, depending on the source and capture method. Liquefaction can be even more energy-intensive, especially in large-scale applications involving transoceanic shipping. Improving the energy efficiency of compression and liquefaction systems is a major focus of current research and development. Innovations such as isothermal compression, advanced materials for heat exchangers, and integrated heat recovery systems are being explored to reduce energy consumption. Additionally, coupling CO_2 processing units with renewable energy sources or waste heat from industrial operations offers opportunities to lower emissions and operating costs. Modular and scalable system designs are also being developed to better suit varying project sizes and geographic conditions.

Applications in CCS Infrastructure

Compressed and liquefied CO_2 technologies are deployed across a range of CCS scenarios. In pipeline-based systems, CO_2 is typically compressed at the capture site and transported to a nearby injection well for underground storage. This method is already in commercial use in regions like North America and Norway. Liquefied CO_2, on the other hand, is essential for maritime transport, enabling international collaboration where captured CO_2 is shipped from industrial regions to offshore storage sites. These technologies also support emerging applications such as direct air capture (DAC) and bioenergy with carbon capture and storage (BECCS). In DAC, where capture is highly distributed and energy-intensive, the ability to liquefy and store CO_2 locally before shipment adds flexibility to deployment. In BECCS, where biomass combustion produces CO_2 that must be stored to achieve negative emissions, reliable compression and liquefaction infrastructure ensures that carbon is sequestered safely and permanently.

Safety and Environmental Considerations

Handling high-pressure gases and cryogenic liquids involves inherent risks, making safety a top priority in system design and operation. CO_2 is not toxic at low concentrations, but it can pose asphyxiation risks in confined spaces. Leakage of compressed or liquefied CO_2 can lead to sudden pressure drops or icing, both of which can damage equipment or endanger personnel. Therefore, facilities must be equipped with gas detection systems, emergency ventilation, and automatic shutdown capabilities. From an environmental standpoint, the CO_2 compression and liquefaction processes do generate indirect emissions, primarily from energy consumption. If this energy is sourced from fossil fuels, it can offset some of the benefits of carbon capture. Therefore, integrating these systems with low-carbon energy sources is essential for maximizing their net climate benefits. Care must also be taken to manage any process by-products or thermal pollution resulting from cooling systems.

3.2 INFRASTRUCTURE CHALLENGES FOR CO_2 TRANSPORT

3.2.1 Safety Considerations in CO_2 Transportation

The transportation of carbon dioxide (CO_2) is a critical component of the carbon capture, utilization, and storage (CCUS) chain, enabling the movement of captured CO_2 from industrial sources to storage or utilization sites. As CCUS deployment expands globally, ensuring the safety of CO_2 transportation has become a top priority for industry stakeholders, policymakers, and the public. Unlike other industrial gases, CO_2 has unique properties that necessitate specific safety protocols and infrastructure standards. Whether transported by pipelines, ships, or trucks, comprehensive safety measures are required to prevent accidents, minimize environmental risks, and protect human health.

The safe transportation of CO_2 is a linchpin in the carbon capture and storage value chain. Whether through pipelines, ships, or other means, it requires a deep understanding of CO_2's physical properties, advanced engineering, strict adherence to safety protocols, and strong institutional oversight. With the right combination of technology, training, and regulation, the risks associated with CO_2 transport can be effectively managed, allowing CCS to play its vital role in global efforts to combat climate change.

CO_2 is an odorless, colorless gas that can pose risks under certain conditions. At high concentrations, it can displace oxygen in the air, leading to asphyxiation. Although it is not flammable or explosive, its behavior under high pressure and low temperature can cause severe physical effects in the event of accidental release. Compressed or liquefied CO_2 is stored at high pressure or cryogenic temperatures, and any sudden decompression or leakage can result in rapid phase changes, cold burns, or pressure surges. These properties demand meticulous design, monitoring, and maintenance of transport systems to ensure integrity and reliability throughout the CO_2 journey.

Pipelines are the most common method for transporting CO_2 overland, particularly for large-scale projects. These pipelines are similar in construction to natural gas pipelines but must be designed to handle the specific thermodynamic characteristics of CO_2, especially when transported in a dense or supercritical phase. Materials used in pipeline construction must be resistant to corrosion, which can be exacerbated by the presence of water or impurities in the CO_2 stream. Therefore, dehydration of CO_2 before transport is a standard safety measure to prevent internal damage to pipelines and associated infrastructure. Additionally, pipeline networks are equipped with pressure control systems, emergency shut-off valves, and continuous monitoring systems to detect any changes in pressure or flow that could indicate a leak or rupture.

Routing and siting of CO_2 pipelines also involve careful planning to avoid densely populated or environmentally sensitive areas. Safety regulations require minimum distances from residential zones and critical infrastructure. Moreover, operators must have emergency response plans in place, including procedures for rapid isolation of pipeline sections and coordination with local emergency services. Public awareness and communication programs are essential components of pipeline safety, ensuring that nearby communities understand the nature of the infrastructure, potential hazards, and steps to take in an emergency.

In addition to pipelines, CO_2 can be transported by ship, especially for international projects or when pipelines are impractical due to geography. Maritime CO_2 transport typically involves liquefied CO_2, which is stored at low temperatures and moderate pressure in cryogenic tanks. Ships designed for CO_2 transport share similarities with liquefied petroleum gas (LPG) carriers, but must accommodate the specific density and pressure-temperature relationships of CO_2. Safety considerations for maritime transport include robust tank insulation, pressure relief systems, and navigation protocols to avoid collision or grounding. During loading and unloading, temperature and pressure must be carefully controlled to avoid thermal shock or over-pressurization.

Port facilities that handle liquefied CO_2 must meet stringent safety standards, including proper containment, spill prevention, and fire protection measures. Personnel involved in CO_2 transfer operations require specialized training to manage the risks associated with cryogenic liquids and pressurized gases. Regular inspections and maintenance of ship tanks, transfer lines, and safety systems help ensure long-term operational integrity. In the event of a leak or accidental release at sea, environmental impact assessments and mitigation protocols must be in place, although CO_2 itself does not pose the same ecological risks as oil or toxic chemicals.

Another mode of CO_2 transport, though less common for large-scale projects, is by truck or rail. These methods are used mainly for short distances or pilot-scale applications. Like ships, trucks and railcars must be equipped with insulated tanks capable of maintaining CO_2 in a liquid or supercritical state. Safety measures include reinforced containment, pressure relief valves, and secure fittings to prevent accidental discharges during transit. Road and rail safety regulations mandate adherence to hazardous material transport protocols, including driver training, routing restrictions, and emergency preparedness.

Monitoring and leak detection are essential elements of CO_2 transportation safety across all modalities. Advances in sensor technology and real-time data transmission enable continuous surveillance of pipeline and vessel conditions. Infrared gas detectors, fiber optic sensors, and acoustic leak detection systems can identify early signs of failure, allowing for timely intervention. Integration of digital technologies such as artificial intelligence and machine learning can enhance predictive maintenance, identifying potential vulnerabilities before they lead to incidents.

Regulatory frameworks play a vital role in enforcing safety standards and building public confidence in CO_2 transport infrastructure. Many countries have developed or adapted regulations to cover the specific needs of CO_2 pipelines and shipping, often modeled on standards used in the natural gas industry. These regulations cover aspects such as pipeline design, construction, operation, maintenance, and decommissioning. International collaboration and harmonization of standards, especially for transboundary CO_2 transport, are increasingly important as global CCS projects expand.

Human factors are also central to the safety of CO_2 transportation. Skilled personnel, comprehensive training programs, and a strong safety culture are necessary to ensure that systems operate as intended. Operators must be equipped to respond effectively to both routine and emergency situations. Simulation-based training, safety drills, and performance audits are common practices to reinforce preparedness and accountability. Transparent reporting of incidents and near-misses helps to continuously improve safety protocols and reduce the likelihood of future accidents.

In the broader context of climate mitigation, safe CO_2 transportation is essential to the credibility and success of CCS initiatives. Accidents or public resistance stemming from safety concerns can delay or derail projects, undermining their contribution to emissions reduction. Therefore, proactive engagement with communities, transparent communication of risks and benefits, and robust safety assurance practices are integral to gaining public trust and regulatory approval.

3.2.2 Cost and Economic Considerations

Carbon dioxide (CO_2) transportation plays an essential role in the carbon capture, utilization, and storage (CCUS) chain, enabling the movement of captured emissions from industrial and energy sources to designated storage or utilization sites. As nations aim to meet their climate targets and transition to low-carbon economies, the scalability of CCUS will largely depend on the availability and economic viability of CO_2 transport infrastructure. However, the development of this infrastructure brings with it a host of financial, logistical, and regulatory challenges that must be thoroughly considered to ensure successful deployment. The development of CO_2 transport infrastructure is vital to unlocking the full potential of carbon capture and storage technologies. However, it comes with significant economic challenges that span capital investment, operational costs, regulatory uncertainty, and public perception. Overcoming these challenges requires a multifaceted strategy involving technological innovation, policy support, financial instruments, and collaborative planning. As governments and industries aim to scale up CCUS as part of their climate strategies, addressing the cost and economic barriers to CO_2 transportation will be crucial for long-term success and sustainability. With the right combination of investment, policy alignment, and infrastructure planning, CO_2 transport can evolve into a reliable backbone of global decarbonization efforts.

Capital Investment and Infrastructure Costs

The cost of constructing CO_2 transport infrastructure is one of the most significant economic barriers to widespread CCUS implementation. Building CO_2 pipelines involves considerable capital expenditure, often higher than those used in natural gas pipelines, due to the need for materials that can resist corrosion and withstand high pressure.

These costs are compounded by necessary investments in compressor stations, safety valves, monitoring systems, and dehydration facilities to ensure the CO_2 is transported in its appropriate supercritical or liquefied state. Additionally, the financial burden of land acquisition, environmental impact assessments, and securing construction permits further elevates the initial investment.

Operational Costs and Maintenance Challenges

Beyond initial construction, the ongoing operational expenses present another layer of economic challenge. Maintaining the pressure and temperature required to transport CO_2 in liquid or supercritical form requires energy-intensive processes, particularly for compression and pumping. Regular maintenance is also critical to avoid leaks and ensure the safety of infrastructure components. These include routine inspections of pipelines, valves, and detection systems. Over time, wear and tear can lead to increased maintenance costs, making operational efficiency a decisive factor in long-term economic sustainability.

Geographical and Spatial Considerations

The cost-effectiveness of CO_2 transport infrastructure is heavily influenced by geographical factors. Many major emission sources, such as power plants and industrial complexes, are located far from suitable geological storage sites. This spatial disconnect requires long-distance pipeline networks or the development of shipping infrastructure, both of which significantly raise the economic stakes. Challenging terrains, urban development, or ecologically sensitive areas can further complicate route planning and increase project costs. As a result, each project must be assessed on a case-by-case basis, accounting for unique geographic and environmental constraints.

Economies of Scale and Network Utilization

The development of large-scale CO_2 transport networks offers the potential for significant cost savings through economies of scale. A shared infrastructure that transports CO_2 from multiple sources to one or more storage sites can reduce per-ton transportation costs and maximize asset utilization. However, achieving these economies of scale requires coordinated planning and long-term commitment from multiple stakeholders. Without guaranteed volumes of CO_2 to transport, underutilization of the infrastructure can lead to financial losses, making the business case for investment less attractive in the absence of a supportive policy environment.

Policy, Incentives, and Financial Risk

Economic uncertainty is further compounded by the lack of well-established market mechanisms and consistent policy frameworks for CO_2 transport. Unlike traditional energy sectors, CO_2 transportation does not yet have a mature commercial structure or clear financial incentives. The absence of strong carbon pricing mechanisms, government subsidies, or tax credits can make it difficult for developers to secure financing. Investors often view CO_2 infrastructure as high-risk due to unclear revenue streams and long payback periods, underscoring the need for government-backed financial instruments such as green bonds, loan guarantees, or direct subsidies to lower investment risk.

Regulatory Hurdles and Legal Uncertainty

Navigating complex regulatory environments adds another dimension to the economic challenges faced by CO_2 transport projects. Variations in regional and national safety standards, classification of CO_2 as a commodity or waste product, and permitting requirements can delay project timelines and increase compliance costs. The lack of legal harmonization for cross-border CO_2 transport also creates uncertainty for projects that involve maritime or international pipeline transport. These regulatory barriers can deter investment, especially if developers are unsure of long-term legal liabilities or the conditions for infrastructure ownership and access.

Public Perception and Social License to Operate

While often overlooked, public acceptance plays a critical role in the economics of CO_2 transport. Projects that face strong community opposition may experience delays, litigation, or increased compensation requirements. Public concerns often center on safety, environmental impact, and property rights. Addressing these issues through engagement strategies, transparent communication, and environmental safeguards is essential for obtaining a social license to operate. However, such efforts also incur costs, both in terms of time and financial resources, and must be factored into the overall economic feasibility of a project.

Technological Innovation and Cost Reduction Potential

Emerging technologies offer opportunities to reduce costs and enhance the efficiency of CO_2 transport. Advances in pipeline materials, corrosion prevention, and real-time leak detection systems can extend asset lifespans and lower maintenance costs. Modular and flexible transport solutions, such as mobile CO_2 containers or integrated pipeline networks, are being explored to adapt to smaller-scale projects or decentralized storage systems. While these innovations show promise, they also require upfront investment and rigorous testing before being adopted at commercial scale, which may limit their immediate impact on economic considerations.

Integration and Infrastructure Synergies

Integrating CO_2 transport with existing infrastructure networks can reduce overall development costs and minimize land-use conflicts. For example, utilizing decommissioned oil and gas pipelines or co-locating new pipelines along existing utility corridors can avoid duplicative permitting and reduce construction costs. Shared use of port facilities, storage tanks, and transportation hubs can further lower capital expenditures for maritime CO_2 shipping. However, achieving such integration depends on multi-stakeholder cooperation, strategic infrastructure planning, and regulatory alignment across sectors, which are not always easy to achieve.

Financing Models and Business Strategies

Innovative financing approaches are crucial to overcoming the economic hurdles of CO_2 transport infrastructure. Public-private partnerships can leverage government support while attracting private investment. Some developers are exploring third-party access models where infrastructure is owned and operated by a neutral entity and used by multiple emitters under standardized contracts. Others advocate for the creation of CCUS hubs or clusters where infrastructure is centrally coordinated, reducing individual costs and risk. Ultimately, the success of these models depends on market demand, policy stability, and the existence of long-term agreements between emitters, transporters, and storage providers.

3.2.3 Regulatory and Environmental Compliance

Regulatory and environmental compliance represents one of the most complex and impactful challenges facing CO_2 transport infrastructure. While these regulations are essential for ensuring safety, environmental protection, and social equity, they can also slow down project deployment and increase financial risk. Overcoming these challenges will require coordinated policy frameworks, stakeholder engagement, technological innovation, and international cooperation. By addressing compliance issues head-on and embedding them into infrastructure planning from the outset, stakeholders can build a robust and trusted CO_2 transport system that supports global decarbonization goals without compromising environmental integrity or public trust.

Fragmented Regulatory Frameworks across Jurisdictions

One of the most significant challenges in CO_2 transport infrastructure development is the fragmented nature of regulatory frameworks across different jurisdictions. National, regional, and local governments may apply different definitions, standards, and procedures for approving, operating, and monitoring CO_2 pipelines and shipping operations. In some countries, CO_2 is still classified as a waste product rather than a commodity, which imposes additional restrictions under hazardous material laws. This lack of harmonization makes it difficult for project developers to plan cross-border or even inter-regional pipelines, often requiring multiple layers of approval and compliance with conflicting regulations. The absence of a clear, unified regulatory pathway adds uncertainty and risk to infrastructure investment decisions.

Permitting and Land Use Authorization Delays

Securing permits and land use rights for CO_2 transport infrastructure can be a lengthy and complex process, often marked by bureaucratic delays and stakeholder opposition. Developers must navigate a maze of environmental regulations, zoning laws, and landowner negotiations before construction can begin. Environmental impact assessments (EIAs), which are mandatory in many jurisdictions, involve extensive studies on the potential effects of pipeline routes or shipping terminals on wildlife, water bodies, soil integrity, and air quality. These assessments require time and resources and can be subject to public scrutiny and legal challenges.

In densely populated or environmentally sensitive areas, opposition from local communities or advocacy groups can delay approvals and necessitate rerouting or redesign, further complicating the project timeline and increasing costs.

Safety Regulations and Technical Standards

Transporting CO_2, particularly in large volumes and at high pressures, involves a range of safety risks that must be mitigated through stringent technical standards and regulatory oversight. CO_2 is non-flammable, but under certain conditions it can cause asphyxiation or corrosion, especially if it contains impurities such as hydrogen sulfide or moisture. Regulatory agencies typically require adherence to strict design codes for pipelines, including specifications for wall thickness, pressure ratings, corrosion-resistant materials, and emergency shut-off systems. Safety standards also dictate the frequency of inspections, maintenance procedures, and leak detection protocols. Compliance with these safety regulations is essential, but it can also raise the cost and complexity of infrastructure projects, particularly when new technologies or unconventional routes are involved.

Cross-Border Transport and International Agreements

For countries that aim to export or import captured CO_2 across borders—whether through pipelines or maritime shipping—international regulatory frameworks become critically important. However, cross-border CO_2 transport is still a relatively new concept in international law, and existing agreements do not always provide clear guidance on liability, environmental responsibility, or dispute resolution.

The London Protocol, which governs marine dumping, was amended to allow transboundary CO_2 transport for sub-seabed storage, but this amendment has not been ratified by all member countries. This creates legal ambiguity for developers seeking to move CO_2 across international waters. Establishing clear bilateral or multilateral agreements is necessary to ensure that CO_2 transport does not violate environmental treaties or trade regulations, but achieving such consensus can be politically and diplomatically challenging.

Monitoring, Reporting, and Verification (MRV) Requirements

An essential aspect of regulatory compliance in CO_2 transport is the implementation of effective monitoring, reporting, and verification (MRV) systems. Regulators require infrastructure operators to track the quantity and quality of CO_2 being transported, detect leaks, and ensure the integrity of pipelines or shipping containers. Advanced sensor technologies, real-time monitoring platforms, and data analytics are increasingly being employed to meet MRV obligations.

However, installing and maintaining these systems adds to the operational burden and costs of compliance. Moreover, data must be reported in standardized formats and submitted to multiple agencies, requiring robust documentation and auditing mechanisms. Failure to comply with MRV standards can result in fines, project suspensions, or reputational damage.

Environmental Justice and Community Consent

In many regions, regulatory agencies are placing greater emphasis on environmental justice and the need for meaningful community engagement. Projects that affect vulnerable or marginalized communities must now demonstrate how they will minimize negative impacts and distribute benefits equitably. This includes addressing concerns related to health risks, property devaluation, and disruptions to local ecosystems.

Gaining community consent requires transparent communication, public consultations, and sometimes compensation or benefit-sharing agreements. Although these measures enhance social legitimacy and reduce the risk of opposition, they also increase the administrative workload and project costs. Furthermore, community consent is not always guaranteed, especially in areas with a history of environmental harm or distrust toward industrial activities.

Climate Regulations and Lifecycle Emissions

Environmental compliance in CO_2 transport also involves aligning projects with broader climate regulations and national emission reduction targets. While transporting CO_2 helps reduce atmospheric emissions, the infrastructure itself consumes energy and can generate emissions during construction, operation, and maintenance. Regulators may require lifecycle assessments to ensure that CO_2 transport projects deliver net climate benefits. These assessments must account for emissions from compressor stations, ships, and ancillary equipment. If the emissions from transport and storage outweigh the captured CO_2, the project may face regulatory disapproval or lose eligibility for carbon credits. Balancing the climate impact of infrastructure operations with their mitigation potential is a critical aspect of environmental compliance.

Insurance, Liability, and Long-Term Responsibility

Infrastructure developers and operators must also address issues of liability and insurance, particularly in the event of accidents, leaks, or environmental damage. Regulatory frameworks often require companies to demonstrate financial capacity to manage such risks, including through insurance policies, contingency funds, or performance bonds. However, the long timeframes involved in CO_2 storage and the relatively new nature of CO_2 transport technologies make it difficult to predict long-term liabilities. Questions arise about who is responsible for monitoring and remediation decades after a pipeline is decommissioned or a shipping route is closed. Without clear legal provisions, uncertainty over liability can deter investment and hinder the growth of the CO_2 transport sector.

Integration with National and Regional Policies

For CO_2 transport infrastructure to gain regulatory approval and environmental clearance, it must be integrated into broader national and regional development policies. This includes alignment with energy transition plans, transportation strategies, and environmental protection programs. Projects that lack alignment with these broader goals may face regulatory pushback or lose access to public funding. Coordinated planning between different government departments and agencies is essential to streamline permitting, avoid duplication of effort, and ensure that infrastructure is developed in a way that supports long-term climate objectives.

Need for Harmonized International Standards

Given the global nature of climate change and the increasing potential for transboundary CO_2 transport, there is a growing need for harmonized international standards and best practices. Organizations such as the International Energy Agency (IEA) and the International Organization for Standardization (ISO) have started developing technical guidelines and protocols for CO_2 transport and storage. However, adoption remains uneven, and many countries still lack domestic legislation that aligns with international norms. Developing universally accepted standards for safety, MRV, and environmental compliance would significantly reduce regulatory complexity and encourage investment in cross-border infrastructure.

3.3 CO_2 NETWORKS: CONNECTING CAPTURE SOURCES TO STORAGE SITES

3.3.1 Hub-and-Spoke Models for CCS Networks

Carbon Capture and Storage (CCS) has become an essential component in the global strategy to mitigate climate change by reducing greenhouse gas emissions from industrial sources and energy generation. One of the primary challenges in large-scale CCS deployment lies in building an efficient and cost-effective infrastructure that connects carbon capture sources to suitable geological storage sites. The hub-and-spoke model has emerged as a promising approach to organizing this infrastructure. In this model, a centralized hub is connected to multiple CO_2 emission sources and storage locations via a network of pipelines or transportation routes. This system allows for the aggregation of CO_2 from various capture facilities and its distribution to one or more storage sites, thereby optimizing logistics, lowering costs, and enhancing scalability. The model mimics successful infrastructure paradigms seen in transportation and logistics sectors and is especially suitable for regional or national CCS strategies. The hub-and-spoke model presents a strategic and efficient blueprint for organizing CCS infrastructure at regional and national levels. By centralizing key functions and enabling shared use of transport and storage assets, this model reduces costs, increases flexibility, and supports the gradual expansion of carbon management systems. Its ability to integrate existing industrial infrastructure, leverage technological innovation, and align with policy objectives makes it particularly suited to accelerating the deployment of CCS. As climate change mitigation efforts intensify, hub-and-spoke CCS networks are likely to play a pivotal role in enabling large-scale carbon reduction while maintaining economic competitiveness and energy security.

The Structure and Function of a Hub-and-Spoke Model

At its core, a hub-and-spoke model in CCS involves the establishment of central hubs where captured CO_2 from multiple sources is collected, processed if necessary, and then transported to storage sites. These hubs serve as convergence points for different emitters, such as power plants, cement factories, chemical refineries, and steel mills, which are typically clustered in industrial zones. From these hubs, CO_2 is transported through pipeline networks or other conveyance systems to designated geological storage locations, such as depleted oil and gas fields, saline aquifers, or offshore reservoirs. The spokes in this model refer to the pipeline connections linking each emission source to the hub and from the hub to the storage endpoint. By centralizing infrastructure at the hub, redundancy is reduced, economies of scale are achieved, and overall project management is streamlined.

Advantages of the Hub-and-Spoke Configuration

The hub-and-spoke model offers numerous advantages that make it an attractive framework for CCS network development. One of the most significant benefits is cost-efficiency. Instead of each individual CO_2 emitter constructing its own pipeline to a storage site—which would be prohibitively expensive and logistically complex—a shared infrastructure at the hub allows multiple users to distribute costs and resources. This significantly reduces capital and operational expenditure. Furthermore, the modular nature of this model allows for gradual scaling. New emission sources can be added to the network by simply connecting additional spokes to the existing hub, and more storage sites can be connected as capacity demands increase. This flexibility makes it possible to align infrastructure development with policy incentives and market readiness.

Geographical Suitability and Industrial Clusters

The success of a hub-and-spoke CCS network largely depends on geographical and industrial factors. Areas with a high density of emission sources, such as industrial clusters or port cities, are especially well-suited for this model. These regions benefit from the proximity of multiple emitters, which makes pipeline routing more efficient and reduces the need for long-distance transport. Additionally, the presence of existing infrastructure such as energy grids, roadways, and industrial utilities can support the development and operation of the CCS hub. In such locations, pre-existing knowledge of subsurface geology and the availability of storage capacity further support the viability of centralized CCS operations. Countries like the Netherlands, the United Kingdom, and Norway have begun to implement hub-and-spoke CCS networks in such industrial regions to maximize impact and minimize costs.

Operational Synergies and Shared Services

The hub-and-spoke model facilitates operational synergies through the centralization of key services such as compression, dehydration, metering, and quality control. Instead of duplicating these systems at every capture site, a single, well-equipped hub can manage the conditioning and dispatch of CO_2 to storage sites. This reduces technical complexity and lowers the chances of operational errors. The central hub can also serve as the control center for monitoring, reporting, and verification (MRV) systems, thereby improving regulatory compliance and safety oversight. Moreover, economies of scale make it feasible to invest in advanced technologies and redundancies at the hub level that would be unaffordable for individual facilities acting alone. Such centralized service provision enhances the overall reliability and sustainability of the CCS network.

Integration with Existing Energy and Pipeline Infrastructure

One of the promising aspects of hub-and-spoke CCS networks is their potential integration with existing energy and pipeline infrastructure. Many regions possess legacy oil and gas pipelines that can be repurposed for CO_2 transport, provided they meet safety and technical specifications. Similarly, depleted hydrocarbon reservoirs and associated injection wells can be adapted to serve as CO_2 storage sites. The hub-and-spoke design allows for the flexible adaptation of these existing assets by connecting them into a centralized network. This not only lowers capital investment but also speeds up project timelines. In some cases, natural gas hubs and energy ports are being converted into CCS hubs, leveraging their strategic locations, workforce expertise, and logistical networks.

Policy and Regulatory Support for Hub-and-Spoke CCS

The development of hub-and-spoke models requires supportive policies, clear regulations, and long-term planning. Governments play a crucial role in facilitating land use approvals, environmental permits, and infrastructure financing. Policy instruments such as tax credits, carbon pricing, and public-private partnerships are instrumental in making CCS networks economically viable. Furthermore, governments can encourage the development of CCS clusters by designating carbon management zones, streamlining permitting processes, and offering incentives for early movers. The regulatory environment must also address the coordination of multiple stakeholders, the ownership and liability of shared infrastructure, and compliance with monitoring and safety standards. Successful CCS hubs often emerge from collaborations between governments, industrial operators, research institutions, and financial stakeholders, highlighting the need for an integrated policy framework.

Technological Innovations Enhancing Network Efficiency

Advancements in technology are making hub-and-spoke CCS networks more effective and adaptable. Innovations in pipeline materials, compression systems, and leak detection sensors are improving safety and reducing maintenance costs. Digital platforms and AI-powered monitoring systems enhance real-time data collection and analysis, ensuring early detection of irregularities and facilitating predictive maintenance. Geographic information systems (GIS) and modeling tools are also aiding in the optimal routing of pipeline networks and the assessment of geological storage sites. These technological improvements help reduce operational risks and increase the attractiveness of CCS networks to investors and regulatory bodies alike.

Case Studies Demonstrating the Model in Action

Several real-world examples demonstrate the practical application of the hub-and-spoke model for CCS. The Northern Lights project in Norway is a leading initiative that collects CO_2 from industrial sites across Europe and transports it via ship and pipeline to an offshore storage site in the North Sea. Similarly, the Porthos project in the Netherlands involves a CO_2 collection hub in the Port of Rotterdam, connected to multiple industrial emitters and a pipeline to offshore geological storage. In the United Kingdom, the East Coast Cluster and HyNet North West are two major initiatives structured around this model, aiming to decarbonize industrial regions by developing coordinated CCS networks. These projects underscore the feasibility and scalability of the hub-and-spoke approach and serve as blueprints for future initiatives worldwide.

Challenges and Considerations in Network Deployment

Despite its benefits, the hub-and-spoke model faces several challenges. Initial capital costs for hub infrastructure can be significant, especially when developing from scratch. Coordinating multiple stakeholders with different financial and operational interests can be complex and time-consuming. Regulatory uncertainties, especially regarding long-term liability and cross-border CO_2 transport, can pose obstacles. Technical challenges such as maintaining CO_2 purity, ensuring pressure stability, and preventing corrosion also need careful management. Furthermore, public perception and community acceptance can influence the success of infrastructure development, especially if projects are located near populated areas or ecologically sensitive zones. Addressing these challenges requires comprehensive planning, stakeholder engagement, and sustained government support.

3.3.2 Cross-Border Transport and International Cooperation

As nations strive to meet ambitious climate goals and transition towards a low-carbon future, Carbon Capture and Storage (CCS) has emerged as a vital tool in the global decarbonization toolkit. A critical dimension of expanding CCS deployment is the development of CO_2 transport infrastructure that not only connects capture sources to nearby storage sites but also enables the cross-border movement of carbon dioxide. Many countries, particularly those with limited or no suitable geological storage options, depend on regional or international cooperation to export captured CO_2 to other nations with established or emerging storage capacity. This necessitates robust international frameworks, technical standardization, legal agreements, and mutual trust. Cross-border CO_2 transport and the international coordination required to implement it play a key role in scaling CCS globally and ensuring equitable participation in climate mitigation efforts. Cross-border CO_2 transport and international cooperation are indispensable components of a comprehensive climate change mitigation strategy. By enabling the transfer of captured CO_2 from emission sources to storage sites across national borders, countries can overcome geographical limitations, optimize infrastructure use, and participate equitably in the global effort to reduce emissions. While challenges related to infrastructure development, regulatory harmonization, and diplomatic coordination remain, the benefits of cooperation far outweigh the risks. Successful examples already in operation or development prove that international CCS networks are not only feasible but also economically and environmentally advantageous. As the world moves toward deep decarbonization and net-zero emissions, cross-border CO_2 transport will play a pivotal role in connecting the dots between ambition and action.

Rationale for Cross-Border CO_2 Transport

Several regions around the world generate significant volumes of carbon dioxide from industrial activities and power generation but lack the geological conditions required for long-term underground CO_2 storage. For such countries, participating in global CCS efforts hinges on the ability to export captured CO_2 to other regions. Cross-border transport becomes a strategic solution to this limitation by creating a more interconnected and flexible network of capture, transport, and storage systems. In addition, international CO_2 transport enables the development of centralized storage hubs in geologically favorable areas, which can accept CO_2 from multiple nations. This enhances the efficiency of CCS networks by consolidating storage operations, optimizing infrastructure investments, and facilitating economies of scale. The ability to move CO_2 across borders is not just a logistical necessity but a powerful enabler of global collaboration on climate action.

Infrastructure Requirements and Transport Modalities

Cross-border CO_2 transport relies on the development of dedicated infrastructure capable of safely and efficiently moving large volumes of carbon dioxide across national boundaries. This can include pipelines, shipping routes, rail networks, and, in some cases, truck transport. Pipelines are the most common method for transporting CO_2 over land and short distances, while maritime shipping is better suited for international transfer between regions separated by water or when storage sites are offshore. Establishing this infrastructure requires coordinated planning and investment across countries, including alignment on technical standards, safety regulations, and environmental safeguards. Ports and transport terminals must be adapted or constructed to handle liquefied or compressed CO_2, and pipeline networks must comply with cross-border engineering and legal standards. Such infrastructure acts as the backbone of a transnational CO_2 network, linking emitters in one country to storage locations in another.

Legal and Regulatory Frameworks

The successful implementation of cross-border CO_2 transport depends on a supportive legal and regulatory environment. This includes both domestic laws and international agreements that authorize, govern, and facilitate the movement of CO_2 across sovereign territories. Key legal considerations include defining the status of CO_2 under trade and environmental law, addressing liability for potential leaks or accidents, and clarifying the responsibilities of transport operators and storage providers. Harmonizing standards for CO_2 purity, pressure, temperature, and monitoring is also essential to ensure compatibility between national systems. International cooperation must be formalized through bilateral or multilateral treaties that outline technical procedures, dispute resolution mechanisms, and protocols for data sharing and verification. These frameworks help build mutual trust among countries and provide legal certainty to investors, developers, and regulators involved in cross-border CCS projects.

Examples of Cross-Border CCS Initiatives

Several cross-border CCS initiatives are already underway or in advanced planning stages, demonstrating the practical potential of international cooperation in CO_2 transport. One of the most prominent examples is the Northern Lights project in Norway, which aims to transport captured CO_2 from industrial emitters in various European countries to offshore storage sites in the Norwegian continental shelf. This project represents a pioneering effort in establishing a transnational CCS value chain and involves collaboration between governments, industries, and private sector entities. Similarly, the Porthos project in the Netherlands envisions receiving CO_2 from neighboring countries for storage in the North Sea. These projects highlight how cross-border partnerships can bridge the gap between capture and storage capabilities and create regional CCS networks that benefit multiple countries.

Economic and Financial Considerations

Developing cross-border CO_2 transport infrastructure is capital-intensive and requires careful economic planning. Countries must consider the costs of constructing and maintaining pipelines, ports, shipping fleets, and storage facilities. These investments are often beyond the capacity of individual companies and may require public funding, international financing mechanisms, or public-private partnerships. Regional economic organizations such as the European Union can play a critical role in mobilizing funds and aligning investment priorities. The cost-effectiveness of cross-border CCS can be enhanced by pooling resources, sharing infrastructure, and coordinating project timelines. Additionally, international CO_2 transport opens opportunities for storage-rich countries to develop new revenue streams by offering CO_2 storage services, potentially creating new carbon markets and stimulating regional economic development. Ensuring fair cost-sharing arrangements and equitable access to CCS infrastructure is vital for maintaining the integrity of international cooperation.

Environmental and Safety Considerations

Transporting CO_2 across borders involves environmental and safety risks that must be carefully managed. Potential hazards include pipeline ruptures, leakage during shipping, and environmental impacts from infrastructure development. To mitigate these risks, comprehensive environmental impact assessments and safety protocols must be enforced across all jurisdictions involved in the transport network. Emergency response systems, monitoring technologies, and regular inspections are essential components of risk management. Moreover, public transparency and stakeholder engagement are crucial in addressing concerns about safety and environmental integrity. International standards developed by organizations such as the International Organization for Standardization (ISO) and guidelines from the Intergovernmental Panel on Climate Change (IPCC) can provide a basis for consistent risk assessment and safety practices across borders. Aligning environmental policies and safety regulations across countries ensures the long-term sustainability and public acceptance of cross-border CCS operations.

Diplomatic and Political Dimensions

Cross-border CO_2 transport is not just a technical or economic issue; it also involves complex diplomatic and political dynamics. Establishing agreements for CO_2 export and import requires sustained diplomatic dialogue, trust-building, and alignment of national interests. Countries must navigate issues such as sovereignty, national security, and geopolitical considerations when entering into CCS partnerships. Historical relationships, regional alliances, and trade dynamics can influence the willingness of countries to engage in CO_2 transport cooperation. Transparent communication, multilateral engagement forums, and regional dialogues help facilitate consensus and address political concerns. Governments must ensure that international cooperation on CCS aligns with broader climate, energy, and foreign policy objectives. By framing CCS as a shared global responsibility, countries can foster a spirit of collaboration and mutual benefit.

CHAPTER

4

Carbon Storage Technologies

4.1 GEOLOGICAL CO_2 STORAGE

4.1.1 Types of Geological Storage Sites: Saline Aquifers, Oil and Gas Fields, Coal Seams

Saline Aquifers

Saline aquifers have emerged as one of the most promising geological formations for carbon dioxide (CO_2) storage in the global effort to combat climate change. These underground formations consist of porous rock saturated with highly saline water that is unfit for human consumption or agricultural use. Their widespread presence, large storage capacity, and suitable geological characteristics make them attractive candidates for the long-term sequestration of CO_2 produced from industrial processes and power generation. Saline aquifers offer a scientifically sound and technically feasible solution for the long-term storage of CO_2. Their abundant availability, large capacity, and natural geological features make them suitable for mitigating carbon emissions at a meaningful scale. While challenges remain in terms of monitoring, risk management, and public perception, ongoing research and successful pilot projects provide a strong foundation for their wider deployment. As the world seeks sustainable solutions to the climate crisis, saline aquifers stand out as a vital tool in the arsenal of carbon management technologies.

Geological Characteristics of Saline Aquifers

Saline aquifers are typically found at depths greater than 800 meters beneath the Earth's surface, where pressure and temperature conditions are sufficient to maintain CO_2 in a supercritical state. In this dense, fluid-like form, CO_2 can be injected more efficiently into the subsurface. The key geological features that determine the suitability of a saline aquifer for CO_2 storage include its porosity, permeability, caprock integrity, and regional tectonic stability. High porosity allows the rock to hold a significant volume of CO_2, while high permeability facilitates the movement of the injected gas throughout the aquifer. The presence of a non-permeable caprock, such as shale or clay, is essential to trap the CO_2 beneath the surface and prevent its upward migration. These aquifers are typically located in sedimentary basins, which have developed over millions of years through the accumulation of sediments. These basins offer vast storage potential and often include multiple layers of rock, allowing for the possibility of stacked storage systems. The presence of natural brine, which is highly saline water, ensures that the aquifer is not used for potable purposes, reducing the risk of competing interests and potential contamination of freshwater sources.

Mechanisms of CO_2 Trapping in Saline Aquifers

Once CO_2 is injected into a saline aquifer, several physical and chemical trapping mechanisms work in tandem to ensure its long-term containment. The first is structural trapping, which occurs when CO_2 accumulates beneath the impermeable caprock in the form of a buoyant plume. Over time, residual trapping takes place as CO_2 becomes immobilized in the pore spaces of the rock due to capillary forces. Solubility trapping involves the gradual dissolution of CO_2 into the saline water, forming a denser solution that sinks deeper into the formation. Lastly, mineral trapping, which is the slowest but most permanent mechanism, results from chemical reactions between the dissolved CO_2 and the surrounding rock minerals, forming stable carbonate compounds. These mechanisms provide multiple layers of security for the storage process. Structural and residual trapping are relatively quick and effective in the short term, while solubility and mineral trapping contribute to the long-term stabilization of CO_2. The interplay of these mechanisms enhances the reliability of saline aquifers as safe geological storage sites and reduces the risk of leakage or environmental harm.

Site Selection and Characterization

Selecting an appropriate saline aquifer for CO_2 storage requires a comprehensive understanding of its geological, hydrological, and geomechanical properties. Detailed site characterization involves seismic surveys, core sampling, and reservoir modeling to evaluate the aquifer's capacity, injectivity, and containment capabilities. The thickness, lateral extent, and depth of the aquifer are assessed to ensure that it can accommodate the desired volume of CO_2. Additionally, the integrity and continuity of the caprock must be confirmed through geological analysis and pressure testing. Hydrodynamic modeling plays a critical role in predicting the behavior of the CO_2 plume over time. This involves simulating the movement of CO_2 within the aquifer and its interaction with brine and rock materials. Such models help in designing the injection strategy, monitoring plan, and risk assessment procedures. Understanding the regional stress field and fault systems is also essential to prevent induced seismicity or fault reactivation, which could compromise the containment of CO_2.

Monitoring and Risk Management

Ensuring the safety and effectiveness of CO_2 storage in saline aquifers depends on continuous monitoring and risk management. Monitoring begins before injection and continues during and after the storage operation to detect any signs of leakage, plume migration, or geochemical changes. Techniques used for monitoring include time-lapse seismic imaging, well logging, pressure and temperature sensors, and tracer studies. Surface monitoring systems are also employed to detect any CO_2 that might escape into the atmosphere or shallow groundwater systems. Risk management involves identifying potential failure scenarios, such as wellbore leakage, caprock breach, or fault activation. Mitigation strategies are developed based on the likelihood and consequences of these events. These strategies include well remediation, pressure management through brine extraction, and real-time data analysis for early warning. Regulatory frameworks and public engagement also play a role in building trust and ensuring the accountability of storage projects.

Environmental and Societal Implications

The use of saline aquifers for CO_2 storage offers significant environmental benefits by reducing greenhouse gas emissions and mitigating the effects of climate change. It enables the continued use of fossil fuels in a transitional period while renewable energy sources are scaled up. However, there are also environmental risks associated with the leakage of CO_2, which could lead to the acidification of groundwater or the release of toxic metals from rock formations. Proper site selection, engineering controls, and monitoring are essential to minimize these risks. From a societal perspective, the deployment of saline aquifer storage projects requires public acceptance and regulatory oversight. Communities near storage sites may have concerns about safety, land use, and environmental impact. Transparent communication, stakeholder involvement, and clear legal frameworks are necessary to address these concerns and build confidence in the technology. International cooperation and knowledge sharing can also help establish best practices and standards for safe and effective CO_2 storage.

Global Potential and Case Studies

Saline aquifers are widely distributed around the world, offering enormous potential for CO_2 storage across different continents. Countries such as the United States, Canada, Norway, and Australia have identified significant storage capacities in their deep saline formations. The Sleipner project in the North Sea, operated by Equinor, is one of the earliest and most successful examples of CO_2 storage in a saline aquifer. Since 1996, it has stored over a million tonnes of CO_2 annually in a deep formation, demonstrating the technical feasibility and safety of the process. In the United States, the Illinois Basin-Decatur Project has successfully injected CO_2 into the Mount Simon Sandstone formation, highlighting the potential of onshore saline aquifers. Similar initiatives are underway in China, India, and the Middle East as part of global efforts to reduce carbon emissions. These projects serve as valuable test beds for refining storage techniques, monitoring systems, and regulatory frameworks.

Future Prospects and Research Needs

As climate change intensifies and global carbon reduction targets become more ambitious, the role of saline aquifers in carbon sequestration is expected to expand. Future research must focus on improving site characterization techniques, enhancing simulation models, and developing more efficient monitoring technologies. Advances in geochemistry and geomechanics can help in understanding long-term interactions between CO_2, brine, and rock. Additionally, integrating saline aquifer storage with carbon capture technologies at industrial sources can create full-scale CCS systems that are economically viable and scalable. Investments in infrastructure, policy support, and international collaboration will be crucial to unlock the full potential of saline aquifers. Developing a network of storage hubs, standardizing safety protocols, and ensuring transparent reporting can accelerate the adoption of this technology. Ultimately, saline aquifers represent a critical component of the global strategy to achieve net-zero emissions and stabilize the climate.

Oil and Gas Fields

Depleted oil and gas fields are increasingly being recognized as ideal geological formations for the long-term storage of carbon dioxide (CO_2), forming a crucial part of the global strategy to mitigate climate change through carbon capture and storage (CCS). These fields, having once been reservoirs for hydrocarbons, are now viewed as viable locations for permanently sequestering anthropogenic CO_2. The geological characteristics that once allowed them to trap oil and gas for millions of years can now be repurposed to contain injected CO_2 safely. The utilization of such fields leverages existing infrastructure, extensive subsurface data, and proven sealing capacity, making them a cost-effective and scientifically reliable option for CO_2 storage.

Geological Features of Oil and Gas Fields

Oil and gas fields typically consist of porous reservoir rocks that have high permeability, allowing the migration and accumulation of hydrocarbons. These reservoir rocks are capped by impermeable layers, known as caprocks, which prevent the upward migration of fluids and gases. This structural integrity, demonstrated over geological time scales, is one of the most compelling reasons for using these fields for CO_2 storage. The knowledge gained from decades of exploration and production enhances the ability to predict how the reservoir will behave when re-injected with fluids, such as supercritical CO_2.Depleted oil and gas fields are usually found at depths exceeding 800 meters, which is ideal for CO_2

storage because the gas remains in a dense, supercritical state under such pressure and temperature conditions. The existing geological traps, such as structural traps formed by folds and faults, and stratigraphic traps created by variations in sedimentary layers, provide natural containment systems. The reservoir rocks, often sandstones or carbonates, have retained hydrocarbons effectively, suggesting that they can similarly retain CO_2. Their known porosity and permeability allow for relatively accurate estimation of how much CO_2 can be stored and how it will migrate through the formation.

Advantages of Using Depleted Oil and Gas Fields

One of the primary advantages of using depleted oil and gas fields for CO_2 storage is the availability of detailed geological and engineering data. Decades of hydrocarbon extraction have yielded comprehensive information on reservoir structure, pressure regimes, fluid behavior, and rock properties. This reduces the uncertainty and cost associated with site characterization and monitoring. The infrastructure, including wells, pipelines, and surface facilities, can often be repurposed for CO_2 injection and monitoring, leading to significant cost savings and faster deployment compared to developing new storage sites from scratch. Another advantage is the proven sealing capacity of the caprock. Since the field has successfully contained oil or gas for millions of years, it provides confidence in its ability to contain CO_2. Additionally, the history of fluid flow in these fields helps in modeling and predicting CO_2 behavior post-injection. Fields that still contain residual hydrocarbons may also offer the possibility of enhanced oil recovery (EOR), where CO_2 injection helps extract additional oil, creating an economic incentive while simultaneously sequestering carbon.

Mechanisms of CO_2 Trapping in Oil and Gas Reservoirs

When CO_2 is injected into a depleted oil or gas reservoir, several trapping mechanisms come into play. Initially, structural trapping confines the injected CO_2 beneath the caprock in much the same way oil or gas was originally trapped. Over time, residual trapping occurs as CO_2 becomes immobilized in the pore spaces of the rock due to capillary forces. Solubility trapping involves the dissolution of CO_2 into any residual water or hydrocarbons present in the formation, which can then migrate slowly downward, enhancing containment. A particularly promising mechanism in carbonate reservoirs is mineral trapping. Here, CO_2 reacts chemically with minerals in the rock to form solid carbonate minerals, thereby converting the gas into a stable, immobile form. Although this process occurs over a long time scale, it represents the most secure form of storage. The simultaneous operation of these mechanisms ensures the long-term containment of CO_2 and reduces the likelihood of leakage to the surface.

Challenges and Considerations

Despite their advantages, the use of depleted oil and gas fields for CO_2 storage also presents certain challenges. The integrity of old wells is a major concern, as improperly sealed or aging wells could become pathways for CO_2 leakage. Many fields have been intensively drilled, and legacy wells may lack the modern sealing and casing technologies required to withstand high-pressure CO_2 injection. Comprehensive well integrity assessments and, where necessary, remediation efforts are essential before initiating injection operations. Another challenge is pressure management. Depleted reservoirs have lower pressures than their original state due to years of production, and re-pressurization through CO_2 injection must be carefully managed to avoid caprock failure or induced seismicity. Modeling pressure buildup and conducting baseline seismic surveys are important steps in ensuring the stability of the storage site. Environmental regulations and public concerns also need to be addressed through transparent risk communication and adherence to rigorous safety standards.

Monitoring and Verification

Monitoring is a critical component of any geological CO_2 storage project, particularly in oil and gas fields where previous activity may have altered subsurface conditions. Baseline data must be established before injection begins, followed by continuous monitoring during and after injection. Techniques such as time-lapse seismic imaging, well logging, surface gas flux measurements, and pressure monitoring provide valuable information about the movement and behavior of CO_2 in the reservoir. Advanced monitoring systems can detect changes in reservoir conditions and alert operators to potential leaks or unanticipated migration. Geochemical sampling of fluids can also provide evidence of CO_2 dissolution and interaction with rock formations. In many cases, satellite monitoring and aerial surveys are used to observe surface changes and ensure environmental safety. Effective monitoring not only safeguards the environment but also builds public and regulatory confidence in the technology.

Economic and Strategic Implications

The economic viability of storing CO_2 in oil and gas fields is enhanced by the potential for cost recovery through enhanced oil recovery operations. EOR involves injecting CO_2 into partially depleted reservoirs to mobilize and extract additional oil, which can make storage projects more financially attractive. In such cases, a portion of the injected CO_2 remains permanently stored in the reservoir, achieving both economic and environmental objectives. Government incentives, carbon pricing mechanisms, and subsidies can further support the financial feasibility of these projects. From a strategic perspective, the re-use of oil and gas infrastructure aligns with the transition toward a low-carbon economy. It provides an opportunity for oil and gas companies to play an active role in climate mitigation while leveraging their technical expertise and asset base. The development of regional CO_2 storage hubs centered around mature oil and gas fields can facilitate the decarbonization of industrial clusters and create new employment opportunities in the energy transition.

Global Examples and Success Stories

Several CO_2 storage projects around the world have successfully utilized depleted oil and gas fields. The Weyburn-Midale project in Canada is a well-known example where CO_2 captured from a coal gasification plant is injected into a mature oil field for EOR and permanent storage. The In Salah project in Algeria, although later discontinued, provided valuable lessons in monitoring and reservoir management. In the United States, the Permian Basin has become a leading site for CO_2-based EOR, with growing emphasis on the storage aspects of these operations. Europe is also investing in this technology, with projects in the North Sea leveraging offshore gas fields for storage. Norway's Snøhvit and Sleipner projects are early demonstrations of offshore geological CO_2 storage, setting benchmarks in safety and effectiveness. These examples illustrate the feasibility of oil and gas fields as storage sites and highlight the importance of adapting techniques to local geological and operational conditions.

Future Outlook

As global climate commitments intensify, the role of oil and gas fields in geological CO_2 storage is expected to grow. Research and development efforts are focusing on improving reservoir modeling, well integrity technologies, and monitoring systems to enhance safety and efficiency. Policy frameworks that support CO_2 storage through legal clarity, financial incentives, and public engagement are essential for accelerating deployment. Partnerships between governments, industry, and academia are also crucial in advancing the knowledge and infrastructure required for successful CCS implementation. Oil and gas fields, once symbols of carbon emission, are now being reimagined as key assets in the transition to a net-zero future. Their geological reliability, economic potential, and technological compatibility make them one of the most practical and scalable options for long-term CO_2 storage. With responsible management and continued innovation, these fields can help bridge the gap between current energy systems and a more sustainable future.

Coal Seams

Coal seams, also known as coal beds, are now gaining increasing attention as potential geological storage sites for carbon dioxide (CO_2). Traditionally known for their role in energy production through coal mining, these subterranean formations have properties that can also be utilized for the permanent sequestration of greenhouse gases. Storing CO_2 in coal seams is a significant component of carbon capture and storage (CCS) strategies aimed at reducing the impact of industrial emissions on global climate. When CO_2 is injected into coal seams, it adsorbs onto the coal's surface, displacing methane gas in a process known as Enhanced Coal Bed Methane (ECBM) recovery. This dual benefit—storing CO_2 while recovering methane—makes coal seams an intriguing and potentially economically viable option for carbon sequestration.

Geological Properties of Coal Seams

Coal seams are composed primarily of organic material that has undergone millions of years of heat and pressure to form solid fossil fuel. These seams are typically located at varying depths underground and are composed of microporous carbon-rich rocks. The physical structure of coal, particularly its cleats and matrix pores, allows for the adsorption of gases such as methane and CO_2. The cleat system consists of natural fractures in the coal that facilitate fluid movement, while the microporous matrix serves as the main gas storage medium. The ability of coal to adsorb gas is a fundamental property that makes it suitable for CO_2 storage. The adsorption process is governed by pressure and temperature conditions, with CO_2 showing a higher affinity for coal surfaces than methane. This means that when CO_2 is injected into a coal seam, it can displace methane, allowing for both carbon storage and gas production. The structural integrity of coal seams, along with their relatively low permeability, helps in containing the injected CO_2, although the permeability must still be sufficient to allow gas injection and movement.

Mechanisms of CO_2 Trapping in Coal Seams

Unlike in saline aquifers or depleted oil and gas fields where CO_2 is primarily trapped through structural and residual mechanisms, in coal seams the primary trapping method is adsorption. When CO_2 is injected, it is physically adsorbed onto the internal surfaces of the coal's micropores. This adsorption is a strong, reversible physical bond that immobilizes the gas without changing its chemical composition. The process is favored by the coal's high surface area and microporosity, which together provide abundant sites for gas adherence. In addition to adsorption, some structural trapping may occur where CO_2 is held in place by caprock layers or within isolated cleat systems. If the CO_2 is injected under certain temperature and pressure conditions, it can also dissolve into formation water, contributing to solubility trapping. Over the long term, chemical interactions may lead to the formation of carbonate minerals, although this form of mineral trapping is less significant in coal seams compared to other storage formations. The multi-mechanism nature of trapping in coal seams adds an extra layer of security to the storage process.

Enhanced Coal Bed Methane Recovery (ECBM)

One of the most notable advantages of CO_2 storage in coal seams is its synergy with Enhanced Coal Bed Methane recovery. During ECBM operations, CO_2 is injected into a coal seam to displace methane that is adsorbed onto the coal. Since CO_2 has a higher adsorption affinity than methane, it preferentially occupies the available sites, pushing methane out into the cleat system where it can be extracted and used as an energy source. This process allows for two key benefits: the recovery of an additional energy resource and the permanent storage of CO_2. It presents a potentially economically favorable model for CCS, especially in regions with abundant unmined coal resources. However, the success of ECBM depends on factors such as coal rank, seam thickness, gas saturation levels, and formation pressure. Not all coal seams are equally suited for ECBM, and extensive site characterization is necessary to determine feasibility.

Site Selection and Suitability

Selecting coal seams as geological storage sites requires a detailed understanding of their geological, geomechanical, and hydrological properties. The depth of the coal seam is critical, as CO_2 must be in a supercritical state for effective storage, which typically requires depths greater than 800 meters. The coal's rank, which ranges from lignite to anthracite, also plays a significant role in determining its adsorption capacity. Higher-rank coals generally have better gas adsorption properties, although permeability may decrease with increasing rank, posing challenges for injectivity. Other important factors include the seam's thickness, areal extent, natural fracture systems, and the presence of impermeable caprocks. The cleat network must be extensive enough to facilitate gas injection and methane recovery, yet not so fractured that it poses a risk of CO_2 leakage. Hydrogeological conditions, such as the presence of formation water and regional pressure gradients, also influence the storage potential and security of a coal seam.

Monitoring and Risk Management

Monitoring is a vital aspect of CO_2 storage in coal seams, given the unique nature of adsorption-based trapping and the presence of gas migration pathways through cleats and faults. Monitoring begins with the establishment of a pre-injection baseline, followed by regular surveillance using geophysical, geochemical, and remote sensing technologies. Pressure and temperature sensors within monitoring wells provide real-time data on reservoir conditions, while gas composition analysis helps verify the displacement of methane by CO_2.Surface monitoring systems are used to detect any potential CO_2 leakage, particularly near old mine shafts, natural faults, or legacy wells that might act as escape routes. Groundwater monitoring is also necessary to ensure that injected CO_2 does not contaminate nearby freshwater aquifers. Risk management strategies must include contingency plans for well remediation, pressure control, and emergency gas venting. Public and regulatory confidence in coal seam storage projects hinges on the robustness of these monitoring and safety protocols.

Environmental and Economic Considerations

The environmental benefits of using coal seams for CO_2 storage lie primarily in the reduction of atmospheric carbon emissions. If combined with ECBM, the process can also reduce reliance on conventional gas production, which often involves more invasive extraction methods. However, there are environmental concerns related to potential leakage, induced seismicity, and changes in subsurface pressure regimes. Methane leakage during extraction is also a risk that must be managed to ensure that the overall climate benefit of the project is not compromised. Economically, the integration of storage with methane production creates a revenue stream that can offset the costs of CO_2 capture and injection. This makes coal seams particularly attractive in coal-rich regions where unmined or marginal seams are abundant. Government policies, subsidies, and carbon pricing mechanisms can further improve the viability of such projects. However, the economic potential varies significantly based on geological conditions and the market value of recovered methane.

Global Potential and Demonstration Projects

Coal seams suitable for CO_2 storage are found in many parts of the world, including the United States, China, India, Australia, and parts of Europe. These countries have large coal reserves and significant industrial CO_2 emissions, making them ideal candidates for coal seam storage projects. Several pilot projects have been conducted to test the feasibility of ECBM and CO_2 storage, such as the Allison Unit project in the U.S. and the Qinshui Basin project in China. These projects have demonstrated the technical feasibility of injecting CO_2 into coal seams and recovering methane, although challenges related to injectivity, permeability loss, and CO_2 breakthrough have been reported. Continued research is needed to improve understanding of the interaction between CO_2 and coal, develop better injection techniques, and optimize production cycles. Long-term monitoring data from these sites are invaluable for refining models and establishing best practices for large-scale deployment.

Future Outlook

As the world seeks scalable solutions to reduce carbon emissions, coal seams represent a unique and underutilized geological resource for CO_2 storage. Their ability to simultaneously sequester CO_2 and produce methane offers a dual benefit that aligns with both environmental and economic goals. However, unlocking this potential will require further research, regulatory support, and technological advancements in reservoir engineering and gas separation. The future of coal seam storage depends on how effectively the challenges of well integrity, permeability loss, and long-term monitoring can be addressed. Advances in geophysical imaging, smart sensors, and predictive modeling will be essential for ensuring the safety and efficiency of these operations. With continued investment and collaboration between governments, industry, and academia, coal seams can become an important part of the global carbon management strategy.

4.1.2 Key Geomechanical Considerations for CO_2 Storage

The secure and long-term geological storage of carbon dioxide (CO_2) requires a comprehensive understanding of geomechanical behavior within the storage site and its surrounding formations. Geomechanics refers to the study of how geological materials deform and react under stress, including changes in pressure, temperature, and chemical interactions. In the context of CO_2 storage, geomechanical considerations are crucial because the injection of CO_2 into subsurface formations alters the natural equilibrium of geological structures. If not properly managed, these changes can compromise the integrity of the storage site, potentially resulting in leakage, caprock failure, or induced seismicity. Therefore, geomechanical assessments are vital for site selection, injection planning, risk mitigation, and long-term monitoring of CO_2 sequestration projects. Geomechanical considerations are foundational to the safe and successful storage of CO_2 in geological formations. From the characterization of stress regimes and fault stability to the integrity of caprocks and wellbores, understanding how the subsurface will respond to CO_2 injection is essential for minimizing risks and maximizing storage performance. Integrating geomechanical assessments with monitoring systems and modeling tools enables the proactive management of potential hazards. As the scale of CO_2 storage projects increases to meet climate goals, the importance of geomechanics in ensuring the reliability and safety of these operations will only continue to grow.

Stress Regimes and Reservoir Stability

The first and foremost geomechanical factor to evaluate in CO_2 storage projects is the in-situ stress regime. Subsurface formations are subject to three principal stresses: vertical stress due to overburden weight, and two horizontal stresses influenced by tectonic forces. The relative magnitudes and orientations of these stresses determine the mechanical behavior of the reservoir rock, the caprock, and any fault systems. Understanding the prevailing stress regime helps predict how these geological units will respond to CO_2 injection, particularly regarding fracture propagation and fault reactivation. During injection, increased pore pressure within the storage formation can reduce the effective stress acting on the rock, potentially triggering geomechanical responses. If the pressure exceeds a critical threshold, it can initiate new fractures or reopen pre-existing ones, which may extend into the caprock or faults. These fractures can become potential pathways for CO_2 leakage. Consequently, one of the primary design goals in CO_2 injection projects is to maintain reservoir pressure below the fracture gradient to ensure mechanical stability and prevent unwanted subsurface movements.

Caprock Integrity and Sealing Capacity

Caprock integrity is essential for the successful containment of injected CO_2. The caprock, typically composed of low-permeability materials such as shale or mudstone, acts as a seal that prevents vertical migration of fluids from the reservoir. From a geomechanical perspective, the caprock must withstand stress alterations caused by CO_2 injection without developing new fractures or becoming reactivated along existing weaknesses. To assess this, geomechanical models simulate the stress changes and mechanical strain the caprock will experience during injection. The sealing capacity of caprocks is influenced by their mechanical properties, including tensile strength, Young's modulus, Poisson's ratio, and fracture toughness. Laboratory experiments on core samples help determine how these properties behave under varying pressures and temperatures. It is also important to evaluate the presence of any pre-existing faults or fractures within the caprock. If these features are critically stressed, even minor increases in pressure could cause them to slip, compromising containment. Maintaining a conservative pressure margin below the caprock's failure envelope is essential for preserving its integrity throughout the operational lifespan of the storage site.

Faults, Fractures, and Induced Seismicity

Faults and fractures are natural structural features in the subsurface that significantly influence geomechanical behavior during CO_2 injection. These discontinuities can either enhance storage security by accommodating pressure or pose risks by serving as conduits for fluid migration. Detailed mapping and characterization of faults are vital components of site assessment. Using seismic imaging and borehole data, engineers assess the orientation, length, displacement, and mechanical properties of faults and fractures. A key concern is fault reactivation, which can occur when increased pore pressure reduces the effective normal stress on a fault plane, potentially inducing slip. If this slip extends to the surface or intersects other formations, it can result in CO_2 leakage or even cause earthquakes. Although most induced seismic events associated with CO_2 storage are minor, they can raise public concern and regulatory scrutiny. To manage this risk, injection pressures must be carefully controlled, and fault slip potential should be continuously evaluated using geomechanical models and real-time monitoring data. Stress path analysis and slip tendency calculations help determine which faults are most susceptible to reactivation under injection conditions.

Reservoir Compaction and Surface Deformation

The physical response of the reservoir to CO_2 injection also plays a critical role in geomechanical planning. As CO_2 enters the reservoir, it displaces formation fluids and increases pore pressure, causing the rock matrix to expand or compact depending on its compressibility. This process can lead to changes in porosity and permeability, influencing both injectivity and storage capacity. If compaction is significant, it can affect the long-term stability of the reservoir and the overlying caprock. Surface deformation may also occur as a result of subsurface changes. Although typically small, uplift or subsidence of the ground surface can be detected using satellite-based interferometric synthetic aperture radar (InSAR) or ground-based tiltmeters. Monitoring such changes provides valuable data about the reservoir's geomechanical behavior and helps verify model predictions. Proper reservoir management, including limiting the rate and volume of CO_2 injection, can reduce the likelihood of excessive deformation and maintain storage integrity.

Thermal Effects and Chemical Interactions

Thermal effects are another important geomechanical consideration, especially when the temperature of the injected CO_2 differs significantly from that of the reservoir. Injecting cold CO_2 into a warm formation can cause thermal contraction of the rock, potentially generating tensile stresses that may induce fractures. Conversely, heating due to exothermic chemical reactions or geothermal gradients may expand the rock, altering its mechanical properties. Geomechanical models must account for these thermal changes to predict their impact on stress distribution and reservoir stability. Chemical interactions between CO_2, formation water, and rock minerals can also alter mechanical behavior over time. CO_2-rich fluids can react with minerals in the reservoir and caprock, leading to processes such as dissolution, precipitation, and mineral replacement. These reactions may weaken or strengthen the rock depending on the mineralogy and reaction pathways. For example, dissolution of carbonate minerals could increase porosity and reduce mechanical strength, while precipitation of new minerals may seal fractures but also cause stress buildup. Coupled hydro-mechanical-chemical models are essential for accurately predicting these effects and their implications for long-term storage.

Wellbore Integrity and Geomechanical Stress

The geomechanical environment also affects wellbore stability and integrity, which are critical for safe CO_2 injection. Wells penetrate through multiple geological layers, and any changes in stress, pressure, or temperature can affect casing, cement, and formation bonding. High injection pressures can cause tensile or shear failure in the wellbore, leading to casing deformation, cement cracking, or debonding. These failures can create pathways for CO_2 leakage along the well. Ensuring wellbore integrity involves proper casing design, cement selection, and placement practices that can withstand expected geomechanical stresses. Mechanical integrity tests and cement bond logs are used to evaluate the quality of well construction. Periodic mechanical integrity testing during the project's lifespan helps detect any issues early. In areas with a history of well failures or complex stress regimes, additional reinforcement or alternative injection strategies may be needed.

Geomechanical Modeling and Simulation Tools

Modern CO_2 storage projects rely heavily on geomechanical modeling tools to simulate stress changes, deformation, and failure mechanisms. These models integrate geological, geophysical, geochemical, and hydrological data to provide a comprehensive view of subsurface behavior. Finite element modeling (FEM) and finite difference modeling (FDM) are commonly used techniques to simulate the response of geological formations to CO_2 injection. Models help in determining safe operating pressure limits, identifying high-risk zones, and planning monitoring strategies. They are calibrated using field data and continuously updated as new information becomes available. Sensitivity analyses and scenario testing allow project teams to anticipate potential problems and devise mitigation measures in advance. These models are crucial for obtaining regulatory approval and ensuring public confidence in the safety and effectiveness of CO_2 storage operations.

Regulatory and Monitoring Implications

Geomechanical considerations directly influence regulatory requirements for CO_2 storage projects. Regulatory bodies often require detailed geomechanical assessments during site selection, permitting, and operational phases. This includes providing evidence that the storage formation can withstand injection without compromising caprock integrity, fault stability, or wellbore integrity. Projects must also include comprehensive monitoring plans that track geomechanical indicators such as pressure, deformation, seismicity, and fluid movement. Real-time data from microseismic monitoring, pressure sensors, and geodetic tools are analyzed to detect early signs of stress changes or geomechanical failure. Adaptive management approaches, which allow for changes in operational procedures based on monitoring results, are becoming standard practice. These efforts ensure that geomechanical risks are effectively managed, reducing the likelihood of environmental harm or project failure.

4.1.3 Site Selection and Characterization

The process of site selection and characterization is central to the successful deployment of geological storage for carbon dioxide (CO_2). Geological storage involves injecting CO_2 into deep underground formations where it can be securely contained for thousands of years. These formations may include saline aquifers, depleted oil and gas reservoirs, unmineable coal seams, and basalt formations. To ensure long-term storage integrity and minimize the risk of CO_2 leakage, selecting the appropriate site and thoroughly characterizing its geological properties is essential. This process involves multiple scientific disciplines, including geology, geophysics, geomechanics, hydrogeology, and geochemistry. The selection and characterization of geological storage sites form the foundation of effective carbon sequestration strategies. This complex, multidisciplinary process ensures that the chosen site can safely and permanently store CO_2, minimizing environmental risks and maximizing public and regulatory confidence. By thoroughly evaluating geological, hydrogeological, geomechanical, and geochemical properties, scientists and engineers can design robust storage systems that play a vital role in mitigating climate change. As carbon capture and storage technologies continue to evolve, the importance of rigorous site selection and characterization will remain central to their success and scalability.

Criteria for Site Selection

The selection of a geological storage site begins with broad regional assessments to identify areas with potential geological formations suitable for CO_2 storage. Key selection criteria include depth, porosity, permeability, caprock integrity, and structural containment. A suitable formation must be deep enough—typically more than 800 meters—to keep CO_2 in a supercritical state, which has favorable density and flow properties for storage. The storage reservoir should possess high porosity to accommodate large volumes of CO_2 and sufficient permeability to allow for effective injection rates without generating excessive pressure build-up. Another vital requirement is the presence of an overlying caprock that acts as a seal to prevent vertical migration of CO_2. The caprock should be thick, laterally extensive, and have low permeability to maintain long-term containment. Additionally, the geological structure of the site, including the presence of faults or fractures, must be stable and well understood. Sites located away from major fault zones are generally preferred to minimize the risk of induced seismicity or leakage pathways. Accessibility, existing infrastructure, and proximity to CO_2 sources also influence site selection, especially in projects aimed at reducing industrial emissions.

Geological Characterization of the Storage Formation

Once a potential site is identified, detailed geological characterization is conducted to understand the physical and chemical properties of the storage formation. This involves both surface and subsurface investigations using data from geological maps, outcrop studies, well logs, and core samples. The objective is to delineate the extent, thickness, and geometry of the reservoir formation and assess its capacity to store CO_2 safely and efficiently. Porosity and permeability are primary characteristics that determine how much CO_2 the formation can hold and how easily it can be injected and distributed. Core sampling and laboratory analyses help determine these properties and provide data on the mineral composition and mechanical strength of the rock. High-resolution seismic surveys are used to create three-dimensional models of the subsurface, allowing for accurate mapping of geological layers, faults, and structural traps. These models are essential for estimating storage capacity, simulating CO_2 flow, and identifying potential leakage pathways.

Caprock Evaluation and Seal Integrity

The effectiveness of CO_2 storage heavily depends on the integrity of the caprock. Caprocks, usually composed of shale, claystone, or evaporites, provide a physical barrier that prevents the upward migration of CO_2. Characterizing the caprock involves analyzing its thickness, mineralogy, mechanical properties, and the presence of any natural discontinuities. Laboratory tests measure capillary entry pressure, which indicates the ability of the caprock to resist fluid invasion. Faults and fractures within the caprock pose significant risks, as they can serve as conduits for CO_2 escape if reactivated or improperly sealed. Therefore, structural mapping and fault analysis are necessary to assess the stability of these features under changing pressure and temperature conditions. Geomechanical modeling is used to predict how the caprock will respond to injection-induced stress and whether it remains within safe limits. Ensuring a robust seal is fundamental to long-term containment and is a critical factor considered during regulatory approval processes.

Hydrogeological and Fluid Flow Analysis

Understanding the hydrogeology of the storage site is equally important in site characterization. Hydrogeological assessments examine the movement of fluids within the formation and surrounding strata. This includes evaluating the hydraulic conductivity, pressure gradients, and the presence of any connected aquifers or faults. If CO_2 were to migrate from the injection zone, the hydrogeological framework determines where and how fast it might travel. Simulation models are employed to forecast the movement of CO_2 through the formation over time. These models require data on fluid properties, formation pressures, and rock permeability to accurately predict plume behavior. One goal of such modeling is to ensure that CO_2 remains within the designated storage complex and does not impact adjacent formations or potable groundwater resources. Baseline hydrogeological data are also essential for future monitoring efforts, allowing for the detection of any anomalies that may indicate leakage or unexpected migration.

Geochemical Interactions and Mineral Trapping Potential

Geochemical analysis is an important aspect of site characterization that focuses on the interactions between CO_2, formation water, and rock minerals. When CO_2 is injected into a geological formation, it dissolves into formation water and can react with minerals to form stable carbonate compounds—a process known as mineral trapping. These reactions contribute to the long-term immobilization of CO_2 and enhance storage security. Geochemical studies involve analyzing formation water samples and rock cores to identify reactive minerals and assess their behavior in the presence of CO_2. Laboratory experiments and reactive transport modeling help predict the rate and extent of mineralization processes. Understanding geochemical dynamics also assists in identifying any potential risks, such as the dissolution of minerals that could increase porosity and weaken rock structure. Geochemical characterization is vital for evaluating storage permanence and designing effective monitoring protocols.

Risk Assessment and Environmental Considerations

Site selection and characterization also involve comprehensive risk assessments to evaluate potential environmental and operational hazards. These include the risk of CO_2 leakage, induced seismicity, groundwater contamination, and wellbore integrity failure. Risk assessments rely on integrated data from geological, geophysical, hydrogeological, and geomechanical studies and are often presented through detailed risk matrices or probability models. Environmental impact assessments are conducted to understand how the storage project might affect nearby ecosystems, land use, and communities. Mitigation strategies are developed to address potential impacts, and regulatory frameworks often require extensive documentation and contingency plans. Public engagement and transparency during this phase are critical to building trust and obtaining a social license to operate.

Infrastructure, Monitoring, and Regulatory Framework

The availability of infrastructure, including access roads, pipelines, and injection wells, plays a practical role in site selection. Proximity to CO_2 sources reduces transportation costs and improves project feasibility. Existing wells can be repurposed for monitoring or injection, but their integrity must be thoroughly evaluated to ensure they do not compromise the storage site. Monitoring systems must be designed during the site characterization phase to track CO_2 behavior during and after injection. This includes installing pressure and temperature sensors, conducting time-lapse seismic surveys, and using satellite-based surface deformation tools.

Monitoring ensures that the CO_2 remains contained and provides data to validate and update predictive models. Regulatory approval for CO_2 storage projects requires detailed documentation of site selection, characterization, and risk assessment activities. Regulations vary by country but generally demand proof of geological suitability, environmental safety, and long-term containment assurance. Compliance with these regulations ensures accountability and helps align the project with national and international climate goals.

4.2 MONITORING, VERIFICATION, AND ACCOUNTING (MVA)

4.2.1 Monitoring CO_2 Plumes and Ensuring Safe Storage

The safe and effective long-term storage of carbon dioxide (CO_2) in geological formations depends on a robust monitoring system that can track the movement and behavior of CO_2 plumes within the subsurface. Monitoring serves multiple purposes: it verifies that the CO_2 remains confined within the intended storage zone, ensures the integrity of geological seals, detects potential leakage pathways, and provides essential data for model validation. Effective monitoring not only helps operators manage storage sites but also builds public and regulatory confidence in carbon capture and storage (CCS) as a climate mitigation strategy. As CO_2 is injected into the subsurface, it displaces formation fluids and begins to migrate through porous rock layers. Monitoring this process involves a combination of geophysical, geochemical, geomechanical, and surface-based technologies designed to offer real-time and long-term information. Monitoring CO_2 plumes and ensuring safe storage are foundational aspects of geological sequestration. Through a combination of advanced subsurface imaging, fluid chemistry analysis, surface detection technologies, and continuous well integrity assessments, operators can confidently manage the movement and containment of CO_2 in deep geological formations. These monitoring systems not only safeguard environmental and public health but also support regulatory compliance and stakeholder confidence. As the deployment of carbon capture and storage expands to meet climate goals, robust and transparent monitoring practices will be indispensable in ensuring the long-term success and credibility of these vital technologies.

Importance of Monitoring in CO_2 Storage Projects

Monitoring plays a crucial role in risk management and regulatory compliance. Geological storage sites are complex, dynamic environments where the behavior of CO_2 must be understood and controlled over decades or even centuries. Without reliable monitoring, it would be impossible to confirm that injected CO_2 remains within the targeted reservoir or to detect early signs of system failure. Regulatory frameworks in many jurisdictions require comprehensive monitoring plans as a condition for project approval. These plans must demonstrate that the operator can effectively track plume migration, assess environmental impacts, and implement corrective measures if necessary. Additionally, monitoring results provide the data needed to improve numerical models, helping to refine predictions and reduce uncertainties in storage performance.

Subsurface Monitoring Techniques

Subsurface monitoring focuses on tracking the CO_2 plume within the reservoir and assessing changes in pressure, temperature, and fluid composition. One of the most widely used techniques is seismic monitoring, particularly time-lapse (4D) seismic surveys. These surveys compare repeated seismic images of the subsurface before and after CO_2 injection to detect changes in acoustic properties caused by the presence of CO_2. Seismic monitoring offers detailed spatial resolution, enabling operators to visualize plume migration and identify any unexpected movement toward faults or caprock boundaries. In addition to seismic methods, downhole monitoring tools such as pressure and temperature gauges, fiber optic sensors, and geophones provide continuous real-time data. These instruments are installed in observation wells surrounding the injection site and measure changes in reservoir conditions as CO_2 is injected. By analyzing this data, operators can assess the pressure front's movement and ensure it remains within safe limits to prevent caprock fracturing or fault reactivation. Distributed Acoustic Sensing (DAS) technology, which uses fiber optic cables to detect vibrations, is increasingly employed to monitor seismic activity and plume dynamics with high sensitivity.

Geochemical Monitoring of Fluids

Geochemical monitoring involves analyzing fluid samples from wells to detect chemical changes associated with CO_2 presence. When CO_2 dissolves in formation water, it lowers the pH and alters the chemical composition of the fluid. Monitoring these changes provides evidence of CO_2 migration and helps verify containment. Key parameters include concentrations of dissolved CO_2, bicarbonate ions, trace metals, and isotopic signatures. By comparing baseline data collected before injection with subsequent measurements, operators can identify any anomalies that might indicate leakage or migration outside the storage zone. Tracer techniques are also used to monitor CO_2 behavior. Tracers are chemical or isotopic substances co-injected with CO_2 and later detected in monitoring wells. They help differentiate between natural background variations and CO_2-induced changes. Tracers can also be tailored to provide information on flow paths, storage efficiency, and breakthrough times. Combined with geochemical data, tracer studies enhance the resolution of subsurface monitoring and offer a powerful tool for verifying storage security.

Surface and Near-Surface Monitoring

Surface and near-surface monitoring techniques are designed to detect any signs of CO_2 leakage that might impact groundwater, soil, or the atmosphere. While geological storage sites are selected and engineered to prevent such leakage, continuous monitoring ensures that any unanticipated migration is quickly identified and addressed. Soil gas monitoring is commonly used to measure CO_2 concentrations at the ground surface. Specialized sensors or sampling probes detect elevated CO_2 levels that may indicate leakage. However, since CO_2 is also produced naturally by biological activity, these measurements must be interpreted in conjunction with isotopic and temporal data to distinguish anthropogenic sources. Atmospheric monitoring involves measuring CO_2 concentrations in the air above the storage site using fixed stations, mobile units, or remote sensing technologies. Instruments such as infrared gas analyzers and laser spectrometers can detect changes in CO_2 levels with high precision. InSAR (Interferometric Synthetic Aperture Radar) is a satellite-based technique used to monitor surface deformation. Uplift or subsidence can result from pressure changes in the subsurface and may indicate injection-related geomechanical effects. Regular InSAR imaging allows for the early detection of such changes, helping operators assess the integrity of the reservoir and surrounding formations.

Microseismic Monitoring and Induced Seismicity

Microseismic monitoring is essential for detecting small-scale seismic events that may be triggered by CO_2 injection. These events, often too small to be felt at the surface, provide valuable information about stress changes, fracture development, and fault activity. Seismometers and geophones placed in wells or on the surface continuously record seismic signals. The location, magnitude, and frequency of these events are analyzed to determine whether the injection process is causing unwanted geomechanical responses. Induced seismicity is a potential risk in CCS operations, particularly if CO_2 injection increases pore pressure near critically stressed faults. Monitoring helps identify the onset of such activity and allows operators to modify injection rates or volumes to reduce the likelihood of fault reactivation. In some cases, a traffic light system is employed, where seismic thresholds are defined, and injection is paused or stopped if certain levels are exceeded. This proactive approach ensures that storage remains safe and does not pose a hazard to surrounding communities.

Wellbore Monitoring and Integrity Assurance

Maintaining the integrity of wells used for CO_2 injection and monitoring is critical to storage safety. CO_2 can be corrosive, particularly in the presence of water, and may degrade well components over time. Wellbore monitoring involves regular inspections using tools such as cement bond logs, ultrasonic scanners, and casing inspection logs. These instruments assess the condition of casing, cement, and formation interfaces to detect any signs of degradation, leakage, or mechanical failure. Continuous pressure and temperature measurements in the wellbore help detect anomalies that could indicate leakage or cross-flow between formations. Monitoring annular pressures and performing periodic integrity tests are standard practices to ensure that the well remains sealed and capable of withstanding operational pressures. In cases where well integrity is compromised, remediation techniques such as re-cementing or casing repair can be implemented to restore safety.

Data Integration and Model Validation

Effective monitoring requires integrating data from multiple sources to create a comprehensive picture of storage performance. Data from seismic surveys, pressure and temperature measurements, fluid analyses, and surface monitoring are combined and compared with predictions from reservoir models. This integration allows for model validation, which is critical for updating simulations and improving forecasts of CO_2 plume behavior. Advanced visualization tools and data analytics platforms assist in processing and interpreting large datasets generated by monitoring systems. Machine learning and artificial intelligence are increasingly being applied to detect patterns, identify anomalies, and optimize monitoring strategies. This continuous feedback loop ensures that the project remains adaptive and responsive to new information, reducing uncertainty and enhancing storage reliability.

Regulatory Frameworks and Public Assurance

Monitoring is a legal and ethical requirement in most jurisdictions overseeing CO_2 storage. Regulatory agencies demand comprehensive monitoring plans that demonstrate the operator's ability to ensure safe and permanent storage. These plans must include details on the types of monitoring technologies used, the frequency of data collection, response protocols in case of leakage, and long-term stewardship responsibilities. Compliance with these requirements is essential for obtaining and maintaining permits, as well as for transitioning responsibility to governmental bodies during the post-closure phase. Beyond regulatory obligations, monitoring also plays a key role in building public trust. Transparent communication of monitoring results and responsiveness to concerns help demonstrate that CCS projects are safe and environmentally responsible. Public engagement initiatives, including informational meetings and access to monitoring data, foster greater understanding and acceptance of CO_2 storage efforts.

4.2.2 Leak Detection Technologies and Techniques

Leak detection is a critical component in the management of geological storage sites for carbon dioxide (CO_2). Ensuring the integrity of these storage sites over long timescales requires comprehensive monitoring systems capable of detecting any unintended migration or leakage of CO_2. Effective leak detection technologies are essential for verifying the containment of CO_2, protecting the surrounding environment, complying with regulatory standards, and maintaining public confidence in carbon capture and storage (CCS) initiatives. The primary goal of these techniques is to provide early warning of any leakage, allowing for prompt remediation and mitigation efforts. Leak detection technologies and techniques are fundamental to the successful implementation and long-term safety of geological CO_2 storage. Through a combination of geophysical imaging, geochemical sampling, atmospheric monitoring, soil gas analysis, and well integrity diagnostics, operators can detect potential leaks early and implement appropriate responses. The integration of these methods into a comprehensive monitoring system ensures that CO_2 remains securely stored and that environmental and public health are protected. As CCS deployment grows globally, the continuous advancement and refinement of leak detection technologies will play a pivotal role in ensuring the success and acceptance of carbon storage as a cornerstone of climate mitigation efforts.

The Importance of Early Leak Detection

The early identification of CO_2 leakage is vital to maintaining the long-term security of geological storage sites. Leaks may occur through faults, fractures, improperly sealed wells, or compromised caprocks. Even small amounts of CO_2 escaping from storage reservoirs can undermine the effectiveness of CCS projects and potentially affect human health, groundwater quality, or surface ecosystems. Early detection allows for the implementation of corrective measures before the leak escalates, reducing environmental damage and financial liabilities. Moreover, continuous and accurate leak detection is fundamental to regulatory compliance, especially in jurisdictions that require long-term post-closure monitoring. Early leak detection also plays a crucial role in validating the storage model and improving future storage site designs.

Geophysical Monitoring Methods

Geophysical methods are among the most widely used technologies for subsurface leak detection. These techniques rely on measuring physical properties of the subsurface that change in response to CO_2 movement. Time-lapse seismic surveys, also known as 4D seismic monitoring, are one of the most effective tools for detecting leakage. By comparing seismic images acquired at different times, operators can detect changes in rock properties caused by the presence of CO_2 outside the storage formation. These surveys provide high-resolution images of the subsurface and can reveal plume migration or unexpected vertical movement toward the caprock. Electrical resistivity tomography (ERT) is another geophysical technique that helps detect leakage. It measures changes in the electrical conductivity of rocks, which can vary when CO_2 displaces brine or other conductive fluids. CO_2 is less conductive than saline water, so its movement alters the electrical response of the formation. Similarly, electromagnetic surveys detect changes in subsurface conductivity and are useful for tracking CO_2 in both saline and dry formations.

Gravity monitoring can also be used in leak detection by identifying subtle changes in gravitational fields resulting from fluid displacement or density variations caused by CO_2 leakage.

Geochemical Detection Techniques

Geochemical methods are essential for identifying chemical changes associated with CO_2 leakage. These changes may occur in soil, groundwater, or subsurface fluids. By sampling and analyzing these media, geochemists can detect elevated CO_2 levels and determine whether they result from natural sources or leakage from storage reservoirs. Key indicators include changes in pH, dissolved inorganic carbon, bicarbonate ions, and trace metal concentrations. One of the most reliable geochemical techniques is the use of isotopic analysis. Carbon isotopes, such as □13C and 14C, help distinguish between biogenic and anthropogenic CO_2 sources. For instance, CO_2 released from deep storage formations often has a different isotopic signature than that produced by microbial activity in the soil. This distinction allows researchers to confirm whether detected CO_2 originates from the storage reservoir. In addition, noble gas tracers like krypton and xenon are sometimes co-injected with CO_2 to act as chemical fingerprints. Their detection in monitoring wells or surface emissions can help confirm a leak and trace its pathway.

Atmospheric and Surface Monitoring Systems

Atmospheric monitoring systems are deployed at the surface to detect any CO_2 escaping from the ground into the air. These systems use a variety of sensors to continuously measure atmospheric CO_2 concentrations, temperature, humidity, and wind patterns. Fixed monitoring stations equipped with infrared gas analyzers or cavity ring-down spectrometers provide real-time measurements and can detect small increases in CO_2 levels. Mobile sensors mounted on drones, vehicles, or handheld devices allow for flexible and rapid assessment of large areas. Remote sensing technologies, such as satellite-based monitoring, can complement ground-based systems by covering extensive geographic regions. Instruments on satellites measure thermal emissions and spectral absorption to detect CO_2 anomalies at the surface. These techniques are especially valuable for monitoring remote or difficult-to-access sites. Additionally, thermal infrared imaging and hyperspectral sensors can detect surface temperature changes or gas plumes that indicate leakage.

Soil Gas Monitoring and Biological Indicators

Soil gas monitoring is a surface technique used to measure CO_2 concentrations in the soil. Sensors or probes are inserted into the ground at various depths to collect gas samples, which are then analyzed for CO_2 and other indicators of leakage. Because soil naturally emits CO_2 from biological activity, baseline measurements are essential for distinguishing between natural and anthropogenic emissions.

When combined with isotopic analysis or tracer detection, soil gas monitoring becomes a powerful tool for confirming the presence and origin of leakage. Biological indicators, though still under investigation, offer a novel approach to leak detection. Changes in microbial communities, vegetation health, or root respiration can signal elevated CO_2 levels in the subsurface. For example, certain plant species exhibit stress symptoms or altered growth patterns when exposed to high concentrations of CO_2. Similarly, microbial populations in the soil may shift in response to chemical changes caused by CO_2. These bio-indicators can serve as supplementary tools for detecting leaks, especially in areas where traditional sensors may be less effective.

Well Integrity Monitoring and Mechanical Diagnostics

Monitoring the integrity of injection and observation wells is a vital aspect of leak detection. Wells can become leakage pathways if their casings, cement seals, or tubing degrade over time. Regular well integrity assessments use tools such as cement bond logs, ultrasonic imaging, and acoustic monitoring to evaluate the condition of well components. These diagnostics can detect casing corrosion, cement channeling, and micro-annuli that could allow CO_2 to escape. Pressure and temperature logging inside the well also helps detect leaks. An unexpected change in pressure or a temperature anomaly can indicate cross-flow between formations or a breach in the well structure. Some sites use fiber optic sensors installed along the wellbore to detect acoustic signals, strain, and temperature changes in real time. These sensors provide continuous data and enable early detection of well integrity issues, allowing for prompt corrective actions.

Integrated Monitoring and Risk Assessment

Given the complexity of geological systems and the diversity of potential leakage scenarios, integrated monitoring approaches are essential. No single technique is sufficient on its own to detect all possible leaks, so combining multiple technologies provides a more comprehensive and reliable detection system. Data from geophysical, geochemical, atmospheric, and well integrity monitoring are synthesized into a unified model of the storage site. This integration allows for cross-validation of results, reduces uncertainty, and enhances the accuracy of leakage detection. Risk assessment models are used in tandem with monitoring data to evaluate the likelihood and impact of potential leaks. These models consider geological characteristics, operational conditions, historical data, and monitoring results to identify high-risk areas and prioritize monitoring efforts. Adaptive monitoring plans are updated regularly based on new information, ensuring that the system remains responsive and effective throughout the lifecycle of the storage site.

Regulatory and Public Confidence Implications

Effective leak detection is not only a technical requirement but also a regulatory necessity. Government agencies mandate the implementation of monitoring systems as part of the permitting and compliance process for CCS projects. Operators must demonstrate that their leak detection technologies can meet regulatory thresholds and that they have contingency plans in place should a leak occur. Compliance with these requirements ensures environmental safety and legal accountability. Moreover, robust leak detection systems are essential for maintaining public trust. Communities near storage sites are often concerned about the potential risks associated with CO_2 storage. Transparent reporting of monitoring results, along with clear communication about the technologies and safeguards in place, helps build stakeholder confidence. Demonstrating the ability to detect and respond to leaks enhances the credibility of CCS as a safe and viable climate solution.

4.2.3 Regulatory and Compliance Standards for MVA

Monitoring, Verification, and Accounting (MVA) form the cornerstone of safe, effective, and transparent carbon dioxide (CO_2) storage in geological formations. As Carbon Capture and Storage (CCS) emerges as a vital climate mitigation technology, regulatory frameworks and compliance standards surrounding MVA become increasingly important. These standards ensure that the injected CO_2 remains contained within the designated storage site, prevent environmental or health risks, and maintain public trust. Regulatory and compliance standards guide operators on how to monitor storage sites, verify that CO_2 has been securely stored, and accurately account for the quantities injected, stored, or potentially emitted. MVA is not just a technical necessity but also a legal obligation under national and international environmental protection protocols. Regulatory and compliance standards for MVA in CO_2 geological storage are essential to ensuring the environmental integrity, operational safety, and public acceptance of carbon capture and storage technologies. These standards provide a structured framework for monitoring plume behavior, verifying storage performance, accounting for CO_2 volumes, and managing long-term responsibilities. By mandating rigorous monitoring protocols, third-party verification, transparent reporting, and post-closure oversight, regulators help mitigate risks and reinforce the credibility of CCS as a climate solution. As the deployment of CCS expands globally, harmonized and robust MVA standards will be critical in maintaining public trust, ensuring environmental protection, and meeting climate goals.

The Role of MVA in CO_2 Geological Storage

The primary purpose of MVA in geological storage is to confirm the long-term containment of CO_2 in deep underground formations and to detect any signs of leakage or migration outside the intended storage zone. Monitoring involves the systematic collection of data on reservoir conditions, CO_2 plume movement, geomechanical stability, and environmental impact. Verification assesses the accuracy of monitoring data and confirms that operations align with predicted models and regulatory expectations. Accounting involves the precise quantification of CO_2 injected, stored, and emitted. Together, these elements help ensure that CCS projects achieve their carbon mitigation goals safely and transparently. The effectiveness of MVA is rooted in its ability to provide defensible and traceable records that regulators, stakeholders, and the public can trust.

Regulatory Frameworks for MVA Implementation

Different countries have developed specific regulatory frameworks for the implementation of MVA in CCS projects. These frameworks define the requirements for site characterization, baseline data collection, monitoring technologies, verification procedures, reporting formats, and post-closure responsibilities. For instance, in the United States, the Environmental Protection Agency (EPA) has established regulations under the Safe Drinking Water Act's Class VI wells for the injection of CO_2 into deep underground formations. These rules require a detailed MVA plan that includes baseline site characterization, real-time monitoring of CO_2 injection, pressure testing, seismic monitoring, and periodic re-evaluation of site models. In the European Union, the CCS Directive (2009/31/EC) provides comprehensive legislation on geological storage, mandating that operators monitor CO_2 plume behavior, reservoir pressure, and leakage pathways using approved techniques. Countries like Australia and Canada have also developed stringent MVA guidelines based on both domestic environmental policies and international agreements. These include project-specific risk assessments, performance-based monitoring protocols, and long-term liability provisions. These regulatory frameworks are underpinned by the principle of adaptive management, requiring MVA programs to be updated regularly based on monitoring results, model refinements, and evolving scientific knowledge. Such flexibility allows operators and regulators to respond proactively to changing site conditions and potential risks.

Standards for Monitoring Techniques and Technologies

To ensure reliability and consistency, regulatory standards specify the types of monitoring technologies that must be employed at various stages of the CO_2 storage lifecycle. During the pre-injection phase, baseline data on geologic formations, groundwater quality, surface gas flux, and seismic activity must be collected to enable meaningful comparison after injection begins. During the operational phase, continuous monitoring is required to track CO_2 plume migration, formation pressure, and caprock integrity. Approved technologies include time-lapse seismic imaging, well logging, microseismic monitoring, pressure and temperature sensors, geochemical sampling, and satellite-based deformation tracking.

Regulators often require that operators use multiple independent monitoring techniques to validate results and reduce uncertainty. For instance, seismic data may be supplemented with geochemical analyses of groundwater to confirm that no migration has occurred. Regulatory standards also emphasize the importance of redundancy, data quality control, calibration of instruments, and long-term data storage. Monitoring systems must remain operational for the duration of the project and often for several decades after injection has ceased, depending on national regulations. The monitoring program must be robust enough to detect anomalies at an early stage, allowing for immediate corrective actions.

Verification Procedures and Third-Party Auditing

Verification is a critical regulatory component that provides assurance that monitoring data are accurate and that the CO_2 is behaving as expected within the geological formation. Regulators require detailed verification protocols that compare observed data with predictive models established during site assessment and permitting. Discrepancies between modeled and observed behavior must be analyzed and explained, and if necessary, the site model must be updated to reflect the new understanding. Third-party verification is often mandated by regulatory frameworks to provide independent assessment of monitoring results and accounting practices. Accredited third-party auditors review MVA reports, inspect monitoring infrastructure, validate data interpretation methodologies, and evaluate the consistency of accounting practices with international standards. Their findings are submitted to regulatory agencies, which may use them to approve continued operations, require corrective actions, or even halt injection if serious deviations are found. The inclusion of independent verification enhances the credibility of MVA systems and reinforces accountability.

CO_2 Accounting and Reporting Requirements

Accurate accounting of CO_2 is essential for regulatory compliance, carbon credit generation, and environmental integrity. Regulatory agencies require operators to quantify the mass of CO_2 captured, transported, injected, and stored with high precision. Emissions from any leakage or venting must also be reported. Accounting must be transparent, traceable, and consistent with international protocols, such as the Intergovernmental Panel on Climate Change (IPCC) Guidelines for Greenhouse Gas Inventories and the ISO 27914 standard on CO_2 storage.

Operators must submit regular reports documenting CO_2 injection rates, cumulative storage volumes, reservoir conditions, and monitoring results. These reports are reviewed by regulatory agencies to ensure that operations remain within permitted parameters and that no unauthorized emissions have occurred. Failure to meet accounting and reporting obligations can result in penalties, revocation of permits, or loss of carbon credits. Accurate CO_2 accounting is also essential for participation in emissions trading schemes or voluntary carbon markets, where stored CO_2 is quantified as a tradable asset.

Post-Closure Monitoring and Long-Term Liability

Regulatory standards also cover the post-closure phase, when CO_2 injection has ceased, but monitoring must continue to confirm the long-term stability of the storage site. Operators are typically required to maintain MVA activities for several decades, depending on the site-specific risk profile and regulatory requirements. Post-closure monitoring ensures that the CO_2 remains confined, that reservoir pressure stabilizes, and that no leakage occurs through legacy wells or geological pathways. Regulatory frameworks define the criteria for site closure and transfer of responsibility from the operator to the government or a designated public authority. This transition usually requires evidence that the storage site is secure, that monitoring results are consistent with predictive models, and that no migration or leakage has occurred. Financial assurances, such as bonds or trust funds, are often required to cover post-closure monitoring and remediation costs. These mechanisms ensure that long-term liabilities are managed responsibly and do not become a burden on taxpayers or future generations.

International Standards and Best Practices

In addition to national regulations, international standards and best practices guide MVA implementation in CCS projects. Organizations such as the International Organization for Standardization (ISO), the Carbon Sequestration Leadership Forum (CSLF), and the International Energy Agency (IEA) provide technical guidance on monitoring design, data management, and reporting protocols. The ISO 27914 standard offers a comprehensive framework for site selection, risk assessment, monitoring methodology, and data verification for CO_2 geological storage. International climate agreements, such as the Paris Agreement, also emphasize the importance of transparent monitoring and reporting in climate mitigation strategies. Countries that include CCS in their nationally determined contributions (NDCs) must demonstrate that storage projects are secure and that emissions reductions are verifiable. Harmonization of MVA standards across countries facilitates cross-border CCS projects and fosters trust in global carbon markets.

4.3 ENHANCING CO_2 STORAGE CAPACITY

4.3.1 CO_2-EOR (Enhanced Oil Recovery) as a Storage Mechanism

Carbon dioxide Enhanced Oil Recovery, or CO_2-EOR, is a process that involves injecting carbon dioxide into oil reservoirs to increase the amount of crude oil that can be extracted. While its primary application has historically been for extending the life of mature oil fields, CO_2-EOR has gained increasing relevance as a method for geological storage of carbon dioxide. It serves the dual purpose of boosting hydrocarbon production and providing a long-term storage option for captured CO_2, thus contributing to efforts to mitigate greenhouse gas emissions. The concept of utilizing depleted or partially depleted oil fields for CO_2 storage provides a technically and economically viable path toward carbon management. CO_2-EOR, therefore, operates at the intersection of fossil fuel recovery and climate change mitigation, presenting unique opportunities and challenges. CO_2-EOR represents a promising convergence of resource recovery and climate mitigation. By injecting carbon dioxide into oil reservoirs, it not only extends the productive life of mature fields but also provides a practical and scalable means of geological CO_2 storage. As part of a broader CCS strategy, CO_2-EOR can help reduce emissions from industrial sources, contribute to negative emissions goals, and support a more sustainable energy system. However, realizing its full potential requires careful reservoir selection, rigorous monitoring, supportive policies, and public engagement. With the right safeguards and incentives, CO_2-EOR can play a vital role in achieving climate targets while ensuring energy security during the transition to a low-carbon future.

Principle and Process of CO_2-EOR

The basic mechanism of CO_2-EOR relies on the miscibility or immiscibility of injected carbon dioxide with the crude oil in the reservoir. In a miscible CO_2-EOR process, the CO_2 mixes with the oil at the molecular level, reducing its viscosity and enabling it to flow more easily toward production wells. This increased mobility allows for the extraction of oil that would otherwise remain trapped in the pores of the reservoir rock. In immiscible processes, CO_2 functions by displacing the oil without fully dissolving into it, still pushing it toward recovery. The CO_2 is typically injected in cycles, alternating with water in a process called water-alternating-gas (WAG) injection, to improve sweep efficiency and maximize oil production. During this process, a portion of the CO_2 remains sequestered in the reservoir, making it a viable option for long-term carbon storage.

Integration with Carbon Capture and Storage (CCS)

Integrating CO_2-EOR with Carbon Capture and Storage (CCS) transforms the technology from an oil production enhancement method into a climate mitigation strategy. Captured CO_2 from industrial sources such as power plants, refineries, and cement factories can be compressed, transported, and injected into oil reservoirs for EOR operations. This integration aligns with global climate goals, as it enables the reduction of CO_2 emissions at source while simultaneously enhancing oil production. Once the CO_2 is injected into the reservoir, a portion of it stays trapped through various mechanisms, including residual trapping, solubility trapping, and mineral trapping, ensuring its long-term containment. This dual benefit increases the economic feasibility of CCS projects, offsetting some of the costs through oil revenues, and thus encouraging broader adoption of carbon management practices.

Reservoir Suitability and Selection Criteria

The suitability of a reservoir for CO_2-EOR and subsequent storage depends on several geological and operational factors. Ideally, the target reservoir should have adequate porosity and permeability to allow CO_2 and oil to move through the rock matrix. It should also possess a competent caprock or sealing formation to prevent upward migration of CO_2 and ensure secure storage. Mature oil fields, especially those with a history of water flooding, are considered prime candidates for CO_2-EOR because their geology is well understood, and infrastructure such as wells and pipelines is often already in place. Detailed reservoir characterization, including pressure, temperature, fluid composition, and structural mapping, is essential to design an effective EOR program and to assess the storage capacity and containment integrity of the reservoir.

CO_2 Retention and Storage Efficiency

One of the critical considerations in using CO_2-EOR as a storage mechanism is the retention rate of the injected CO_2. During the active phase of EOR, a portion of the CO_2 returns with the produced oil and is usually separated, recompressed, and re-injected in a closed-loop system. However, a significant fraction of the injected CO_2 remains permanently trapped in the reservoir. Studies and field data have shown that approximately 30 to 70 percent of the injected CO_2 can be retained in the reservoir depending on the specific geological conditions and the design of the EOR operation. Storage efficiency increases over time as the project transitions from an oil production focus to a storage-oriented phase. Once oil production ceases, operators may continue injecting CO_2 to maximize storage, thereby converting the EOR site into a dedicated storage facility.

Monitoring and Verification Requirements

Effective monitoring and verification are essential for ensuring the secure storage of CO_2 in EOR projects. Monitoring programs must be designed to track the movement of CO_2 in the subsurface, verify its containment within the reservoir, and detect any potential leakage pathways. Techniques such as time-lapse seismic surveys, pressure and temperature logging, tracer injection, and soil gas sampling are commonly employed to gather data on CO_2 behavior. These methods provide critical insights into plume migration, reservoir pressure changes, and caprock integrity. Additionally, operators are required to report injection volumes, production rates, and CO_2 retention metrics to regulatory authorities. The monitoring strategy should also extend into the post-closure phase to confirm long-term containment and fulfill legal and environmental obligations. Ensuring rigorous monitoring builds public confidence and supports the credibility of CO_2-EOR as a legitimate climate mitigation tool.

Environmental and Economic Considerations

While CO_2-EOR contributes to carbon storage, it also presents certain environmental and economic considerations. On the environmental front, concerns have been raised about the potential for induced seismicity, groundwater contamination, and surface infrastructure impacts. These risks must be assessed and mitigated through comprehensive site assessments, engineering controls, and regulatory compliance. Economically, CO_2-EOR offers a revenue stream through enhanced oil production, which can help offset the costs of capturing and transporting CO_2. However, the profitability of such projects is sensitive to oil prices, carbon pricing mechanisms, and the availability of incentives such as tax credits or subsidies for carbon storage. For example, in the United States, the Section 45Q tax credit has provided a significant boost to CO_2-EOR projects by offering financial rewards for every metric ton of CO_2 securely stored. Balancing the economic and environmental dimensions is key to making CO_2-EOR a sustainable part of the energy transition.

Regulatory Frameworks and Policy Support

The deployment of CO_2-EOR as a storage mechanism requires a robust regulatory framework that addresses both hydrocarbon production and carbon storage objectives. Regulations must cover site permitting, operational standards, monitoring requirements, liability provisions, and post-closure responsibilities. In many jurisdictions, CO_2-EOR projects fall under existing oil and gas regulations, which may not fully address the nuances of long-term carbon storage. Therefore, updated or hybrid regulatory approaches are needed to bridge this gap. In addition to national policies, international climate agreements recognize CO_2-EOR as a valid carbon mitigation strategy, provided that the storage of CO_2 is verifiable and permanent. Policy instruments such as carbon markets, emissions trading schemes, and storage certification programs can incentivize the adoption of CO_2-EOR and integrate it into broader climate strategies.

Public Perception and Stakeholder Engagement

Public perception plays a significant role in the deployment of CO_2-EOR projects. While the use of CO_2 for oil recovery may raise questions about continued fossil fuel reliance, it is important to communicate the climate benefits of permanent CO_2 storage and reduced emissions from industrial sources. Transparent communication, stakeholder engagement, and community involvement are essential to address concerns and build support for CO_2-EOR initiatives. Educational outreach, public hearings, and open access to monitoring data can enhance trust and foster a more informed dialogue around the role of CO_2-EOR in climate mitigation.

Technological Advancements and Future Outlook

Technological advancements continue to enhance the efficiency and reliability of CO_2-EOR as a storage mechanism. Innovations in reservoir modeling, real-time monitoring, enhanced injectivity techniques, and advanced separation technologies are improving both oil recovery rates and CO_2 retention. Research is also being directed toward optimizing CO_2-EOR in unconventional reservoirs, including tight formations and shale plays. As carbon capture technologies become more widespread and carbon pricing mechanisms gain traction, the potential for CO_2-EOR to scale up as a climate solution grows. The future of CO_2-EOR will likely involve a transition toward more storage-centric models, where the primary objective shifts from maximizing oil output to maximizing secure CO_2 containment.

4.3.2 Mineral Carbonation and Other Long-Term Storage Methods

The need for long-term and secure storage of carbon dioxide (CO_2) has become more pressing in the context of global climate change and the increasing deployment of carbon capture and storage (CCS) technologies. While subsurface geological formations such as saline aquifers and depleted oil and gas fields have traditionally been favored for CO_2 storage, there is growing interest in alternative methods that offer permanent and stable sequestration. Among these, mineral carbonation stands out for its ability to transform CO_2 into solid, thermodynamically stable minerals. Alongside mineral carbonation, other emerging long-term storage methods, such as ocean-based storage, utilization in durable products, and innovative synthetic approaches, offer complementary solutions that expand the portfolio of carbon storage options. These methods are essential to achieving net-zero emissions and addressing the challenges of long-term climate stabilization. Mineral carbonation and other long-term CO_2 storage methods represent promising and essential components of a comprehensive strategy to combat climate change. By converting CO_2 into stable, inert forms, these methods offer durable and secure sequestration that can operate independently or in conjunction with other carbon capture systems. Although challenges remain in terms of cost, scalability, and technical feasibility, ongoing research and pilot projects are paving the way for broader implementation. As part of an integrated approach to carbon management, mineral carbonation and related technologies have the potential to deliver permanent climate benefits while supporting innovation, resource utilization, and environmental stewardship.

Mineral Carbonation

Mineral carbonation is a natural and engineered process in which CO_2 reacts with metal oxides, particularly those found in silicate and carbonate rocks, to form solid carbonate minerals such as calcite, magnesite, and siderite. This process mimics the natural weathering of rocks but accelerates it through industrial means. The reaction between CO_2 and minerals like olivine (magnesium silicate) or serpentine results in stable carbonate compounds that lock the carbon away permanently. This method of storage is considered extremely secure because the resulting carbonates are resistant to leaching and degradation, thus eliminating the risks of leakage or re-release of CO_2 into the atmosphere. Mineral carbonation can be carried out in situ or ex situ. In situ carbonation involves injecting CO_2 directly into ultramafic or basaltic rock formations underground, where it reacts with naturally occurring minerals. Ex situ carbonation, on the other hand, occurs above ground in a controlled environment, where CO_2 is reacted with crushed rock or industrial waste materials such as slag or fly ash. Both approaches have their merits and limitations, depending on the availability of suitable materials, energy requirements, and local geological conditions.

Advantages of Mineral Carbonation for CO_2 Storage

One of the most compelling advantages of mineral carbonation is its ability to provide permanent and irreversible storage of CO_2. Unlike storage in porous geological formations, where the gas remains in a supercritical state and must be monitored for decades, mineral carbonation transforms CO_2 into a solid form that is inherently stable over geological time scales. This eliminates the need for long-term monitoring and liability management, making it an attractive option for policymakers and project developers. Additionally, mineral carbonation does not require the presence of a confining caprock or the management of reservoir pressure, which simplifies site selection and reduces the risk of leakage. It also offers the potential to utilize abundant and widely distributed ultramafic rocks, including those found in regions that may not have access to deep geological storage formations. Moreover, the use of industrial by-products in ex situ carbonation provides an opportunity for waste valorization and contributes to a circular economy approach.

Challenges in Implementing Mineral Carbonation

Despite its advantages, mineral carbonation faces several technical and economic challenges that must be addressed to enable large-scale deployment. One major limitation is the slow natural reaction kinetics of CO_2 with silicate minerals, which necessitates significant energy inputs to accelerate the process. For ex situ carbonation, this typically involves mining, crushing, and grinding rock materials to increase their surface area, followed by high-pressure and high-temperature reactions. These energy requirements can offset some of the climate benefits unless they are supplied from renewable sources. In situ mineral carbonation, while less energy-intensive, depends heavily on the permeability of the host rock and the availability of reactive minerals. Injecting CO_2 into dense ultramafic rocks can be technically difficult due to low porosity and potential for clogging by precipitated carbonates. Furthermore, scaling up mineral carbonation processes to gigaton levels of CO_2 storage remains a significant logistical and engineering challenge. Developing efficient and cost-effective process technologies, optimizing reaction conditions, and ensuring environmental compatibility are key research priorities in this field.

Emerging Research and Pilot Projects

Several pilot projects and research initiatives around the world are exploring the potential of mineral carbonation for carbon storage. One of the most notable examples is the CarbFix project in Iceland, which has successfully demonstrated in situ mineral carbonation in basalt formations. In this project, CO_2 captured from a geothermal power plant is dissolved in water and injected into basaltic rock layers, where it reacts with calcium, magnesium, and iron to form solid carbonates within a matter of months. The success of CarbFix has provided valuable insights into the feasibility and scalability of this approach. Other research efforts are focusing on developing new catalysts and solvents to speed up the carbonation reaction, as well as optimizing the use of industrial waste materials like steel slag and cement kiln dust. These materials often contain reactive oxides and can be readily sourced near urban or industrial centers, making them ideal candidates for ex situ carbonation. Advances in mineral processing and reactor design also hold promise for reducing the energy and cost barriers associated with this technology.

Ocean-Based and Other Alternative Long-Term Storage Options

In addition to mineral carbonation, various other long-term CO_2 storage methods are being investigated. One such approach is ocean-based storage, which involves the direct injection of CO_2 into the deep ocean or its conversion into stable bicarbonate ions through reaction with seawater. The ocean acts as a vast carbon sink, and theoretically, it has the capacity to store large quantities of anthropogenic CO_2. However, concerns over ocean acidification, ecological impacts, and the risk of CO_2 re-release have limited the practical application of this method. Regulatory and public acceptance challenges also persist, particularly in light of international agreements that seek to protect marine ecosystems. Another emerging avenue is the development of synthetic mineralization processes, such as the electrochemical conversion of CO_2 into stable compounds using renewable electricity. These processes offer the advantage of modularity and can be integrated with renewable energy sources to create negative emissions technologies. Additionally, research is ongoing into the incorporation of CO_2 into building materials such as concrete, aggregates, and carbon-storing polymers. These applications not only sequester carbon in long-lived products but also provide economic value and stimulate industrial innovation. Biochar production, which involves the pyrolysis of biomass to create a stable carbon-rich material, is another long-term storage method under investigation. When applied to soils, biochar can enhance soil fertility and water retention while storing carbon for centuries. This approach combines climate mitigation with agricultural co-benefits, although its scalability and carbon permanence vary with feedstock and environmental conditions.

Policy Support and Future Outlook

The successful deployment of mineral carbonation and other long-term CO_2 storage methods will depend heavily on policy support, research funding, and market incentives. Governments can play a pivotal role by establishing clear regulatory frameworks, supporting pilot projects, and integrating these technologies into national climate strategies. Carbon pricing, tax incentives, and carbon credits for mineralized CO_2 can create financial incentives that drive innovation and investment. International cooperation and standardization of measurement, reporting, and verification (MRV) protocols will also be critical to ensure the environmental integrity of storage projects. Public awareness and acceptance are essential, particularly for technologies involving the subsurface or oceanic environments. Transparent communication and stakeholder engagement will help build trust and overcome potential resistance. In the long term, mineral carbonation and other alternative storage methods are expected to complement conventional geological storage and form an integral part of the global carbon management toolkit. As the urgency of climate action grows, diversifying storage options will be necessary to meet emission reduction targets and stabilize atmospheric CO_2 concentrations. Continued advancements in science, engineering, and policy will shape the role of these technologies in the transition to a sustainable, low-carbon future.

4.4 OCEAN STORAGE OF CO_2

4.4.1 Oceanic CO_2 Sequestration Potential

As the world grapples with the escalating threat of climate change, carbon dioxide (CO_2) sequestration has emerged as a vital strategy to reduce atmospheric greenhouse gas concentrations. Among the various carbon sequestration approaches, the ocean offers one of the largest natural sinks for atmospheric CO_2. Oceans already absorb nearly one-quarter of anthropogenic CO_2 emissions each year, highlighting their immense buffering capacity in the global carbon cycle. However, beyond this natural absorption, researchers are exploring the possibility of enhancing oceanic CO_2 sequestration through engineered methods. These techniques aim to accelerate and increase the ocean's ability to store carbon either by biological or chemical processes. The ocean's vast volume, depth, and chemical composition make it a promising candidate for long-term CO_2 storage, though concerns about environmental impacts, reversibility, and ethical considerations continue to fuel global debates. Oceanic CO_2 sequestration presents both immense opportunities and significant challenges in the global effort to mitigate climate change. The ocean's vast capacity for carbon storage, if carefully harnessed, could play a pivotal role in achieving net-zero emissions. However, the ecological, regulatory, and ethical implications of engineered sequestration methods demand rigorous scrutiny and responsible governance. With continued research, transparent decision-making, and adherence to environmental safeguards, oceanic CO_2 sequestration could become a viable component of the broader climate solution framework, contributing to the preservation of both atmospheric stability and marine ecosystem integrity.

Natural Oceanic Carbon Sequestration Mechanisms

The ocean sequesters carbon through two principal natural mechanisms: the solubility pump and the biological pump. The solubility pump involves the physical absorption of CO_2 at the ocean surface, particularly in colder regions where CO_2 solubility is highest. As surface waters cool, they become denser and sink, carrying dissolved CO_2 into the deep ocean where it can remain isolated from the atmosphere for centuries or even millennia. This process plays a significant role in moderating global CO_2 levels. The biological pump, on the other hand, is driven by marine organisms. Phytoplankton absorb CO_2 during photosynthesis, forming the base of the oceanic food web. When these organisms die or are consumed, the carbon in their bodies may sink to deeper waters or settle on the seafloor, effectively removing CO_2 from the surface ocean and atmosphere. Both natural mechanisms contribute significantly to long-term carbon storage, but they operate at scales and speeds that are insufficient to offset current CO_2 emissions from fossil fuel use.

Artificial Enhancement of Oceanic CO_2 Sequestration

To accelerate oceanic carbon sequestration, scientists are exploring various engineered approaches that mimic or amplify these natural processes. One such method is ocean fertilization, which involves adding nutrients like iron to surface waters to stimulate phytoplankton growth. The goal is to increase biological productivity and thus enhance carbon uptake via photosynthesis. As the phytoplankton die and sink, the carbon within them is transported to deeper layers of the ocean, where it may be stored for long periods.

Another technique under investigation is the direct injection of liquid CO_2 into the deep ocean. This method involves capturing CO_2 from industrial sources, transporting it to coastal or offshore locations, and injecting it at depths greater than 1000 meters. At such depths, high pressure and low temperature help keep CO_2 in a dense phase, either as a liquid or solid hydrate, which can remain on the ocean floor or dissolve slowly in surrounding water. While this approach holds promise for large-scale storage, concerns about localized acidification, marine ecosystem disruption, and the potential for CO_2 leakage have limited its deployment.

Alkalinity enhancement, or ocean liming, is another innovative approach. By adding alkaline substances such as lime or crushed silicate rocks to seawater, this method aims to increase the ocean's capacity to absorb CO_2 from the atmosphere. The added alkalinity shifts the ocean's carbonate equilibrium, allowing more CO_2 to dissolve and be stored as bicarbonate and carbonate ions. This process not only enhances CO_2 uptake but also helps counteract ocean acidification. However, the practical challenges of sourcing, grinding, and distributing large quantities of minerals across vast oceanic regions present significant logistical and economic hurdles.

Long-Term Storage and Stability of Oceanic CO_2

The long-term effectiveness of oceanic CO_2 sequestration depends heavily on the residence time of carbon in the deep ocean. Naturally sequestered CO_2 can remain isolated for hundreds to thousands of years, depending on ocean circulation patterns and mixing rates. For artificially sequestered CO_2, similar timeframes are assumed, especially when injected at great depths or stored in forms that resist dissolution and re-emission. However, unlike geological sequestration in rock formations, oceanic carbon storage is inherently more dynamic and susceptible to disturbances. Changes in ocean temperature, circulation, or biological activity can potentially re-release stored CO_2 to the surface. Moreover, deep ocean storage sites are difficult to monitor and regulate, raising concerns about accountability and verification. The permanence of oceanic storage is thus not guaranteed, and long-term monitoring and modeling are essential to evaluate potential risks and benefits.

Environmental and Ecological Considerations

One of the primary concerns regarding engineered oceanic CO_2 sequestration is its potential impact on marine ecosystems. For instance, ocean fertilization can lead to unintended consequences such as harmful algal blooms, oxygen depletion in deep waters, and alterations in food web dynamics. The addition of nutrients may favor certain species over others, disrupting ecological balances and affecting biodiversity.

Similarly, direct injection of CO_2 into the deep ocean can result in localized acidification, posing risks to deep-sea organisms that are not adapted to rapid pH changes. While some species may tolerate elevated CO_2 concentrations, others, particularly calcifying organisms such as corals, mollusks, and certain plankton, may experience impaired growth and reproduction. These ecological risks must be carefully assessed before implementing large-scale sequestration projects.

Alkalinity enhancement, although potentially beneficial in combating acidification, also raises concerns about the source and environmental footprint of the materials used. Mining and transporting large volumes of alkaline substances could have significant terrestrial impacts, and the consequences of altering ocean chemistry at scale remain poorly understood. Therefore, environmental impact assessments, ecological modeling, and precautionary principles must guide the development of ocean-based sequestration strategies.

Regulatory and Ethical Dimensions

The deployment of oceanic CO_2 sequestration methods also entails complex regulatory and ethical issues. Currently, international frameworks such as the London Protocol and the Convention on Biological Diversity impose restrictions on activities that may harm the marine environment, including large-scale ocean fertilization. These agreements reflect the global consensus that geoengineering efforts in the ocean must be approached with caution and transparency.

Ethically, the idea of deliberately altering ocean systems to store carbon raises questions about human intervention in natural processes, intergenerational equity, and potential moral hazard. Critics argue that focusing on technological fixes may divert attention from the need to reduce emissions at their source. Proponents, however, contend that ocean-based sequestration could be an essential tool in a portfolio of climate mitigation strategies, particularly if global emissions continue to exceed safe thresholds.

Public perception and stakeholder engagement are critical in shaping the acceptability and governance of these technologies. Clear communication of risks, benefits, and uncertainties, as well as inclusive decision-making processes, can help ensure that ocean sequestration projects align with societal values and sustainability goals.

Research Priorities and Future Prospects

To fully assess the feasibility of oceanic CO_2 sequestration, continued research and development are imperative. Key areas of investigation include improving the efficiency and predictability of ocean fertilization, refining models of deep ocean CO_2 behavior, understanding the ecological impacts of altered ocean chemistry, and developing low-impact methods for alkalinity enhancement. Field trials, laboratory experiments, and global collaborations can provide critical data to inform policy and practice. Investment in monitoring technologies, such as autonomous underwater vehicles, chemical sensors, and satellite observations, will be essential for verifying sequestration outcomes and ensuring environmental safety. Interdisciplinary research that combines oceanography, chemistry, ecology, engineering, and social science will also be crucial to designing responsible and effective sequestration strategies.

The future of oceanic CO_2 sequestration depends on a balanced approach that acknowledges both its vast potential and its inherent uncertainties. As climate models project the need for gigaton-scale carbon removal by mid-century, the ocean may offer a significant but not standalone solution. It will likely function best as part of a diversified carbon management strategy that includes terrestrial sequestration, mineralization, reforestation, and emission reduction.

4.4.2 Challenges in Deep Ocean Storage

Deep ocean storage has emerged as a promising method for long-term carbon dioxide (CO_2) sequestration in response to the global climate crisis. As atmospheric concentrations of CO_2 continue to rise, scientists and policymakers are exploring various techniques to reduce greenhouse gas emissions and mitigate their environmental impact. Among these methods, deep ocean storage proposes the injection of captured CO_2 into the ocean's depths, typically below 1,000 meters, where pressure is high and temperatures are low enough to maintain CO_2 in a dense phase. The rationale behind this approach is that the ocean, with its vast volume and capacity, can act as a massive sink, isolating CO_2 from the atmosphere for centuries or even millennia.

While deep ocean storage offers theoretical potential as a long-term carbon sequestration strategy, its implementation is fraught with significant technical, environmental, regulatory, and ethical challenges. The deep ocean is a complex and delicate system, and any attempt to manipulate its carbon dynamics must be approached with caution, transparency, and a firm commitment to environmental stewardship. As research continues and technology advances, deep ocean storage may one day play a role in the broader portfolio of climate mitigation strategies. However, until the associated risks are better understood and addressed, its practical deployment will remain limited and controversial.

Technical and Engineering Complexities

One of the most fundamental challenges in deep ocean storage lies in the technical and engineering requirements for transporting and injecting CO_2 into the deep sea. The logistics of capturing CO_2 from industrial sources, compressing it to a supercritical state, and transporting it over potentially long distances to marine injection sites require a well-integrated infrastructure. Subsea pipelines, specialized ships, or floating platforms must be designed to operate in harsh oceanic environments. These systems must withstand high pressure, low temperatures, and corrosive seawater, making engineering design and maintenance particularly complex and expensive.

Additionally, injecting CO_2 at depths greater than 1,000 meters necessitates precise control over injection rates, locations, and dispersal mechanisms to avoid unwanted side effects. Ensuring the CO_2 remains in a stable phase and does not resurface or rapidly dissolve in the water column requires advanced technology and continuous monitoring. The placement and design of diffusers or injectors must take into account local oceanography, including currents, temperature gradients, and bathymetry, which vary significantly across regions. Achieving reliable, long-term containment in such a dynamic and remote environment is a significant barrier that has yet to be overcome on a large scale.

Uncertain Storage Permanence

Another major concern associated with deep ocean storage is the uncertainty regarding the permanence of CO_2 sequestration. Unlike geological storage in rock formations where CO_2 can be trapped in pores and mineralized, ocean storage relies primarily on the physical separation of CO_2 from the atmosphere. Although injected CO_2 can remain in the deep ocean for hundreds to thousands of years due to slow mixing rates, ocean currents and vertical upwelling could eventually bring stored CO_2 back to the surface. This delayed return could result in a future increase in atmospheric CO_2 levels, particularly if global warming accelerates ocean stratification and circulation changes. The long residence time of deep waters varies depending on location. In some regions, such as the North Atlantic, deep waters may return to the surface within a few centuries, while in others, like the North Pacific, the timescales may be longer. Predicting these movements requires complex oceanographic modeling and a comprehensive understanding of thermohaline circulation, which is currently subject to considerable scientific uncertainty. This lack of predictability undermines confidence in deep ocean storage as a truly permanent solution to carbon emissions.

Ecological and Environmental Impacts

The potential environmental consequences of injecting CO_2 into the deep ocean represent one of the most controversial aspects of this storage method. Direct injection can lead to localized acidification of seawater, as CO_2 dissolves to form carbonic acid, which lowers pH levels. Deep-sea organisms, particularly those that rely on calcium carbonate for shell formation, such as certain mollusks and plankton, are highly sensitive to changes in pH. Lower pH can impair biological functions, reduce survival rates, and lead to significant alterations in species composition and biodiversity.

In addition to acidification, the accumulation of CO_2 in bottom waters could result in increased CO_2 concentrations over large areas, potentially disrupting entire ecosystems. Changes in chemical composition may affect microbial communities and biochemical cycles, including those related to nitrogen and sulfur. The long-term consequences of such disturbances are poorly understood due to limited data on deep-sea ecosystems and their resilience. Furthermore, there is concern that deep ocean CO_2 injection could have cascading effects on the marine food web, carbon cycling, and ocean health.

Another significant environmental risk is the possibility of CO_2 leakage or uncontrolled dispersal. Should the CO_2 not remain at the intended depth, or if it migrates unexpectedly due to currents or system failures, it could have wider ecological implications. Given that the deep sea is among the least studied and most sensitive parts of the biosphere, the margin for error is extremely small, and the potential damage could be irreversible.

Monitoring and Verification Challenges

Ensuring the integrity and performance of deep ocean CO_2 storage over long periods is a daunting task. Monitoring such projects requires the deployment of sensors and observation tools across large and remote areas of the ocean floor. These instruments must be capable of measuring changes in pH, dissolved CO_2 concentrations, temperature, and other indicators over long durations with high precision and reliability. Current monitoring technologies are limited by high costs, technical constraints, and environmental variability. Deep-sea deployment, maintenance, and data retrieval are logistically complex and resource-intensive. Moreover, distinguishing between anthropogenic CO_2 and natural oceanic carbon fluxes adds another layer of complexity. Verifying that stored CO_2 remains in place and does not pose a threat to the marine environment or re-enter the atmosphere is essential for public confidence, regulatory compliance, and the issuance of carbon credits. However, the lack of robust monitoring infrastructure and methodologies hampers the ability to track and certify long-term storage outcomes effectively.

Legal and Regulatory Barriers

The legal and regulatory framework surrounding deep ocean CO_2 storage is still evolving and presents numerous challenges. International conventions such as the London Protocol and the United Nations Convention on the Law of the Sea (UNCLOS) impose strict regulations on ocean dumping and the protection of the marine environment. While some amendments have been proposed to allow carbon storage under certain conditions, many jurisdictions still prohibit or tightly restrict the deliberate disposal of CO_2 into the ocean. There is also the issue of jurisdictional boundaries. Identifying who has the authority to approve, regulate, and monitor deep ocean storage projects, especially those conducted in international waters, is a legal grey area. Establishing liability in case of environmental harm, leakage, or unintended consequences is particularly challenging in such a context. Furthermore, the lack of standardized protocols and procedures for site selection, risk assessment, and long-term responsibility complicates the development of consistent and enforceable regulations.

Ethical and Social Considerations

Beyond the technical and regulatory hurdles, deep ocean storage also raises ethical questions about humanity's role in modifying natural systems on a planetary scale. Critics argue that injecting CO_2 into the ocean could be seen as an act of pollution rather than mitigation, particularly given the uncertainties and potential risks involved. There is concern that reliance on ocean storage might serve as a technological distraction, diverting attention from the urgent need to reduce fossil fuel consumption and transition to cleaner energy sources. Social acceptance of deep ocean storage is limited, partly due to the invisibility and inaccessibility of the deep sea, which makes it difficult for the public to grasp the full implications of such interventions. Engaging stakeholders, including indigenous communities, environmental groups, and the broader public, in discussions about ocean-based climate solutions is essential to ensuring transparency and accountability. Ethical considerations around intergenerational equity also come into play, as the consequences of today's actions could be borne by future generations.

Economic Viability and Cost Considerations

Finally, the economic feasibility of deep ocean storage is a significant concern. The capital and operational costs associated with CO_2 capture, transport, and deep-sea injection are considerable. Without strong financial incentives, carbon pricing, or regulatory mandates, private sector investment in deep ocean storage is unlikely to materialize. Competing sequestration methods, such as geological storage or natural climate solutions like afforestation, may offer more cost-effective and socially acceptable alternatives. In the absence of clear economic drivers, subsidies, or carbon market mechanisms, deep ocean storage may remain a theoretical option rather than a practical solution. Developing the necessary infrastructure, risk mitigation strategies, and long-term governance frameworks will require substantial public investment and international cooperation.

4.4.3 Environmental Impacts and Risks

The pursuit of solutions to mitigate climate change has driven the development of various technologies, including carbon capture and storage (CCS). As part of this strategy, geological and oceanic carbon storage methods have gained attention for their potential to reduce greenhouse gas concentrations in the atmosphere.

However, alongside their promise, these approaches raise serious concerns about environmental impacts and associated risks. It is essential to examine and understand the environmental consequences of implementing large-scale carbon storage techniques in order to ensure their safety, sustainability, and public acceptance. Environmental impacts can arise from the entire lifecycle of a carbon storage project, including site selection, injection, long-term storage, and potential leakage. These impacts may affect ecosystems, water resources, biodiversity, air quality, and even human health. The environmental impacts and risks of carbon storage are diverse, complex, and context-dependent. While carbon storage has the potential to contribute significantly to climate change mitigation, it is not without consequences. From groundwater contamination and ecosystem disruption to induced seismicity and long-term monitoring challenges, each stage of a carbon storage project introduces specific environmental considerations that must be addressed. Safeguarding the environment requires a comprehensive approach that includes rigorous site characterization, robust monitoring systems, adaptive management practices, transparent governance, and stakeholder involvement. Only by acknowledging and managing these environmental risks can carbon storage become a safe, effective, and publicly acceptable component of global climate policy.

Subsurface Environmental Disturbance

One of the primary environmental risks of geological carbon storage is the potential disturbance to subsurface geological formations. Injection of carbon dioxide into deep underground reservoirs such as saline aquifers, depleted oil and gas fields, or unmineable coal seams involves introducing pressurized fluids into rock formations. This process can alter the geomechanical equilibrium of the reservoir, potentially triggering induced seismicity or micro-earthquakes. In cases where faults or fractures are present, pressure buildup may reactivate these structures, leading to the migration of CO_2 beyond the intended storage zone. The integrity of caprock, which acts as a sealing layer to prevent the upward movement of CO_2, is critical to safe storage. If the caprock is compromised due to geochemical reactions, mechanical failure, or pre-existing faults, there is a risk of CO_2 leakage into overlying formations or even the atmosphere. Such leakage not only undermines the effectiveness of the storage but also poses risks to groundwater and soil quality.

Groundwater Contamination

Another significant environmental concern associated with carbon storage is the potential contamination of underground sources of drinking water. If CO_2 migrates from the storage reservoir into shallow aquifers, it can alter the chemical composition of groundwater. When CO_2 dissolves in water, it forms carbonic acid, which can lower pH levels and increase the solubility of heavy metals and other contaminants that are naturally present in the rock matrix. This may result in the mobilization of toxic elements such as arsenic, lead, or cadmium, thereby compromising water quality. Furthermore, the injection of CO_2 into subsurface formations can displace brine or saline water, causing it to migrate into freshwater aquifers. This phenomenon, known as brine displacement, can lead to salinization of groundwater resources, making them unsuitable for agricultural, industrial, or domestic use. Ensuring the isolation of storage zones from overlying freshwater systems is therefore a critical element of site selection and regulatory compliance.

Ecosystem Disruption in Oceanic Storage

For oceanic storage options, particularly the direct injection of CO_2 into deep waters or ocean floor sediments, the environmental risks are pronounced. The deep ocean is a largely unexplored and sensitive ecosystem, hosting unique biodiversity adapted to high pressure, low temperature, and chemical stability. Injecting CO_2 into these regions can lead to localized acidification, altering pH levels and disrupting the physiology of marine organisms. Acidification affects calcifying species such as corals, shellfish, and certain plankton by reducing their ability to form calcium carbonate shells or skeletons. This not only impacts individual species but can also have cascading effects on marine food webs and ecosystem services. Additionally, the introduction of CO_2 may change the composition of microbial communities, which play a vital role in nutrient cycling and carbon processing in the ocean. Long-term environmental consequences are difficult to predict due to the limited understanding of deep-sea ecosystems and their complex interactions.

Surface and Atmospheric Impacts

While carbon storage aims to reduce atmospheric CO_2, accidental releases or surface leaks can result in localized atmospheric concentrations that pose risks to human health and safety. In the event of a significant surface release, such as through abandoned wells, faults, or ruptured pipelines, CO_2 can accumulate in low-lying areas due to its density, displacing oxygen and creating asphyxiation hazards. Although such incidents are rare, they underscore the importance of continuous monitoring and emergency preparedness in carbon storage projects. In addition to safety concerns, surface infrastructure for CO_2 capture, transport, and injection may have environmental impacts related to land use, air quality, and noise. The construction and operation of pipelines, compressor stations, and injection wells can lead to habitat fragmentation, soil compaction, and emission of pollutants from machinery and transport vehicles. These localized environmental impacts must be considered alongside the broader climate benefits of the project.

Risks from Monitoring and Maintenance Failure

The long-term success of carbon storage projects depends heavily on effective monitoring, reporting, and verification systems. However, monitoring failures or inadequate maintenance can increase the risk of undetected leaks and environmental damage. For example, if monitoring equipment fails to detect pressure anomalies or chemical changes in surrounding formations, early warning signs of leakage may go unnoticed until significant damage has occurred. Moreover, the long-term stewardship of storage sites, which may require monitoring for decades or centuries, raises questions about institutional responsibility and financial liability. If monitoring lapses occur due to funding cuts, corporate bankruptcies, or shifts in political will, the ability to manage environmental risks may be compromised. Ensuring robust legal and financial frameworks to support long-term site management is therefore essential for minimizing environmental harm.

Climate Feedback and Uncertainty

While CCS technologies aim to mitigate climate change, the potential for unintended climate feedback effects remains a subject of concern. For example, leakage of CO_2 from multiple storage sites could gradually offset the climate benefits of sequestration efforts. The interaction between stored CO_2 and other greenhouse gases or feedback mechanisms in the atmosphere and biosphere is not fully understood, creating uncertainty about the net impact on global warming. Additionally, relying heavily on carbon storage as a mitigation strategy could delay investments in renewable energy, energy efficiency, and behavioral change. This overreliance on technological solutions may contribute to a phenomenon known as moral hazard, where the perceived availability of future remedies reduces the urgency to reduce emissions today. If storage fails to deliver the anticipated environmental benefits or is abandoned due to environmental concerns, the global community may find itself further behind in meeting climate goals.

Biodiversity and Land Use Implications

The environmental footprint of carbon storage projects extends beyond the storage site itself. Capturing CO_2 from power plants or industrial facilities may require additional energy, leading to increased resource consumption and environmental degradation elsewhere. Land use for infrastructure development can lead to the destruction or alteration of habitats, potentially threatening local biodiversity and ecosystem functions. In regions where land availability is limited or already under pressure from agriculture, urbanization, or conservation needs, the introduction of carbon storage infrastructure may exacerbate competition and environmental stress. Balancing the land use needs of CCS with those of other societal priorities is a complex challenge that requires integrated planning and environmental impact assessments.

Public Perception and Environmental Justice

Public concerns about the environmental impacts of carbon storage projects are often heightened in communities that perceive themselves as vulnerable or marginalized. Issues of environmental justice arise when storage sites are disproportionately located near low-income or indigenous populations who may lack the political power to influence decision-making. These communities may bear the environmental risks and burdens without receiving proportional benefits, such as job creation or improved infrastructure. Furthermore, public opposition based on environmental concerns can delay or derail projects, emphasizing the need for transparent communication, community engagement, and equitable risk-sharing. Incorporating local knowledge, respecting cultural values, and ensuring fair compensation and participation are critical to addressing the broader environmental justice dimensions of carbon storage.

CHAPTER 5

Economic and Policy Considerations in CCS

5.1 THE COST OF CCS TECHNOLOGIES

5.1.1 Capital and Operational Costs of CCS Facilities

Carbon Capture and Storage (CCS) is a critical technology in the portfolio of solutions aimed at mitigating global climate change. It involves capturing carbon dioxide emissions from sources such as power plants and industrial facilities, transporting it to a storage site, and injecting it deep underground for long-term sequestration. While the environmental and climate benefits of CCS are widely recognized, the economic aspect remains one of the most significant barriers to its widespread adoption. Understanding the capital and operational costs associated with CCS facilities is essential for assessing its feasibility, promoting investment, and formulating effective policy frameworks. These costs vary depending on the technology employed, the nature of the emission source, infrastructure requirements, and the location and characteristics of the storage site. The capital and operational costs of CCS facilities are significant and remain one of the major hurdles to widespread deployment. From high initial investments in capture, transport, and storage infrastructure to the ongoing costs of energy consumption, maintenance, and monitoring, CCS presents a complex economic challenge. However, with technological advances, economies of scale, supportive policy frameworks, and strategic planning, these costs can be managed and reduced over time. As the world moves toward deep decarbonization, understanding and addressing the financial aspects of CCS will be crucial for ensuring its role in a sustainable and resilient energy future.

Capital Investment Requirements

The capital costs of CCS projects are primarily incurred during the design, construction, and commissioning phases. These costs are substantial and include expenses for capturing equipment, compression systems, transportation infrastructure, and storage facilities. The capture component alone is often the most expensive part of a CCS project, accounting for roughly 60 to 70 percent of the total capital expenditure. This includes the installation of scrubbers, solvents, or membranes that can efficiently separate CO_2 from flue gases or industrial streams. Different capture technologies—such as post-combustion, pre-combustion, and oxy-fuel combustion—have different capital requirements, with post-combustion generally being the most expensive due to the need to retrofit existing facilities.

In addition to capture systems, CO_2 must be compressed to a supercritical state to facilitate efficient transport. Compressors and associated cooling systems add further capital burden, requiring specialized materials and engineering to handle high pressures and prevent corrosion. The transportation phase typically involves the construction of pipelines, which may stretch for hundreds of kilometers if storage sites are not nearby. The capital costs here include land acquisition, right-of-way clearances, construction materials, safety systems, and pump stations. In some cases, especially offshore projects, marine transportation using ships may be more feasible, albeit at higher initial costs.

Storage facilities also contribute significantly to capital expenses. These include the drilling of injection wells, development of monitoring systems, and site characterization through geological surveys and seismic testing. Saline aquifers, depleted oil and gas fields, and unmineable coal seams each require specific preparation, with varying drilling depths, reservoir assessments, and well completions. Costs rise further when redundancy is built into the system for safety and long-term reliability. Overall, the capital intensity of CCS projects is often compared to that of nuclear power plants, underscoring the financial magnitude of initial investments.

Operational and Maintenance Costs

In addition to high capital costs, CCS facilities incur ongoing operational and maintenance (O&M) costs that must be managed over the life of the project. These costs include labor, utilities, maintenance of equipment, and replacement of consumables, regulatory compliance, and monitoring. The energy penalty associated with CCS is one of the largest contributors to operational costs. Capturing and compressing CO_2 requires a significant amount of energy, which in turn reduces the net efficiency of power plants and industrial facilities. This energy demand translates into higher fuel consumption, increased emissions from the facility itself, and elevated operational expenditures.

The maintenance of capture systems is particularly demanding. Solvent-based capture technologies, for example, require regular replacement or regeneration of solvents, corrosion control measures, and frequent inspection of heat exchangers and absorber columns. Membrane and cryogenic systems also require precision maintenance to ensure consistent separation performance. In addition, compressors, pumps, and auxiliary equipment must operate continuously and efficiently to maintain the required pressure and flow rates for CO_2 transport.

Transport infrastructure, especially pipelines, also requires ongoing maintenance to ensure structural integrity and prevent leaks. Regular inspection using smart pigs, cathodic protection systems, and pressure monitoring is essential. For marine transport, operational costs include fuel, staffing, port fees, and maintenance of cryogenic storage systems onboard ships. These operational costs are compounded by the need for continuous safety measures, including leak detection, emergency shutdown systems, and compliance with environmental regulations.

Storage sites must also be maintained over the long term to ensure the integrity of injected CO_2. This includes monitoring well pressures, groundwater quality, seismic activity, and surface emissions. Technologies such as 3D seismic surveys, borehole logging, and surface gas flux measurements are deployed routinely, and these require trained personnel, data analysis, and equipment maintenance. Over time, these monitoring efforts may be scaled down, but they remain an important part of the operational cost structure, especially for sites subject to stringent regulatory scrutiny.

Economies of Scale and Cost Reduction Potential

While the initial costs of CCS are high, there is potential for cost reductions as the technology matures and deployment scales up. Economies of scale play a critical role in reducing unit costs, particularly in the capture and transport phases. Larger facilities can distribute fixed capital costs over greater volumes of CO_2, improving cost efficiency. Additionally, as more projects are implemented, lessons learned and innovations can lead to improved designs, better integration, and more effective operation strategies. The development of regional CO_2 transport and storage hubs can further reduce costs by sharing infrastructure among multiple emitters.

Research and development efforts are also focused on improving the efficiency and lowering the cost of capture technologies. New solvent formulations, membrane materials, and adsorption techniques are being explored to reduce energy penalties and equipment size. Modular construction and standardized designs can shorten construction times and reduce labor costs. Advances in drilling technologies and subsurface imaging can also bring down the cost of site characterization and well completion.

Policy support in the form of subsidies, tax incentives, and carbon pricing mechanisms can help bridge the cost gap and attract private investment. For example, in jurisdictions where a price is placed on carbon emissions, CCS becomes more economically viable as it allows companies to avoid paying carbon taxes or meet emission reduction obligations. Public funding for early demonstration projects can also de-risk investment and accelerate cost learning across the industry.

Comparative Costs with Other Mitigation Options

The cost of CCS must be evaluated in comparison with other greenhouse gas mitigation options. In many cases, renewables such as solar and wind, energy efficiency improvements, and behavioral changes offer lower-cost alternatives. However, CCS holds a unique position for sectors that are difficult to decarbonize, such as cement, steel, and chemical manufacturing. In these industries, process emissions are inherent to the production method, and CCS may be one of the few available options to achieve deep decarbonization.

For fossil fuel-based power generation, the competitiveness of CCS depends on several factors, including fuel prices, regulatory pressures, and the availability of alternative energy sources. Retrofitting existing power plants with CCS is generally more expensive than building new low-carbon generation facilities, but it may be justified in regions with abundant fossil resources or existing infrastructure. In some cases, CCS can be combined with enhanced oil recovery (EOR), where captured CO_2 is used to increase oil production from mature fields. This can offset some of the costs, although it introduces debates about the net climate benefit.

Financial Risks and Investment Barriers

Financing CCS projects remains challenging due to high upfront capital requirements, uncertain returns, and long payback periods. Private investors are often hesitant to fund CCS without government support or guaranteed revenue streams. The risks associated with technological complexity, regulatory uncertainty, and long-term liability further complicate financing. Lenders and insurers must also consider the potential for environmental damage and remediation costs in the event of a leak or operational failure. Innovative financing mechanisms, such as public-private partnerships, green bonds, and carbon contracts for difference, are being explored to make CCS projects more attractive. Governments can play a catalytic role by providing loan guarantees, grants, or infrastructure investments that reduce financial risk and signal long-term commitment to CCS deployment.

5.1.2 Cost Reduction Strategies and Technological Advancements

Cost reduction strategies and technological advancements are essential to unlocking the full potential of Carbon Capture and Storage as a climate mitigation solution. By investing in more efficient capture technologies, optimizing transport and storage infrastructure, promoting modular system design, and leveraging supportive policy frameworks, stakeholders can significantly lower the economic barriers to CCS deployment. Technological innovation, when combined with strategic planning and robust policy support, creates a pathway for CCS to transition from a niche solution to a mainstream component of global climate strategy. As costs continue to fall and experience accumulates, CCS will become increasingly accessible, scalable, and impactful in the global effort to reduce greenhouse gas emissions.

Advancements in CO_2 Capture Technologies

The CO_2 capture phase represents the most expensive segment of a CCS system, accounting for a significant portion of both capital and operational expenditures. Traditional solvent-based post-combustion capture, which uses amines such as monoethanolamine (MEA), is effective but energy-intensive and costly to maintain. New solvent formulations are being developed to enhance absorption capacity, reduce regeneration energy, and increase thermal stability. These advanced solvents also have lower degradation rates, thereby minimizing replacement frequency and maintenance costs.

In addition to improved solvents, membrane-based separation technologies have made significant progress. Membranes offer a compact, modular, and potentially lower-cost alternative to chemical absorption, especially when integrated with pre-combustion systems or used in combination with other capture methods. Advanced materials such as polymeric, ceramic, and mixed matrix membranes are being optimized for high selectivity and permeability, enabling efficient CO_2 separation with reduced energy consumption.

Cryogenic separation and adsorption-based techniques are also under active investigation. These methods, which rely on cooling and pressure-swing adsorption respectively, are becoming more attractive as their efficiency improves. Solid sorbents, including metal-organic frameworks (MOFs) and zeolites, show promise due to their high surface area and tunable chemistry, which allow for selective and reversible CO_2 capture. The ongoing development and commercialization of these alternative technologies could dramatically reduce capture costs, particularly for smaller or decentralized emission sources.

Process Integration and Optimization

Beyond developing new technologies, integrating CO_2 capture systems into industrial processes in a more holistic manner can lead to significant cost savings. Process integration involves optimizing heat exchange networks, utilizing waste heat from industrial operations to power solvent regeneration, and reducing energy losses through advanced controls and automation. For example, integrating CCS into a power plant's heat recovery system can reduce the parasitic load and improve overall energy efficiency. In industrial facilities, coupling CCS with existing gas treatment units or chemical processes can reduce redundancy and capital requirements. Co-locating capture systems with hydrogen production units or using captured CO_2 as a feedstock for chemicals and fuels can also create synergies that improve economic performance. Digital technologies such as process simulation, digital twins, and machine learning algorithms are increasingly being used to model and optimize CCS systems in real time, allowing operators to identify inefficiencies and proactively address maintenance needs, thereby reducing operational costs.

Transport and Storage Infrastructure Optimization

The development of CO_2 transport and storage infrastructure is another area where cost reduction strategies can be effectively implemented. Building shared CO_2 transport networks, such as regional pipeline corridors or shipping hubs, allows multiple emitters to use the same infrastructure, thereby distributing capital and operational costs across a larger base. This hub-and-cluster approach is being piloted in several regions around the world, particularly in areas with dense industrial activity and suitable storage geology.

Standardizing pipeline design, construction, and monitoring protocols can also drive down costs through mass production and economies of scale. Advances in materials science, such as corrosion-resistant alloys and coatings, can extend pipeline life and reduce maintenance expenses. In marine transport, improvements in liquefaction and storage systems, as well as more fuel-efficient vessels, contribute to the economic feasibility of shipping CO_2 to offshore storage sites.

For the storage phase, the use of enhanced oil recovery (EOR) can provide economic returns that partially offset the costs of capture and transport. Though not a permanent solution, EOR allows operators to monetize captured CO_2 by increasing hydrocarbon recovery. In saline aquifer storage and other long-term sequestration options, advancements in site characterization and reservoir modeling help reduce uncertainty, lower monitoring requirements, and improve injection efficiency, all of which contribute to cost reduction.

Modular and Scalable System Designs

One of the emerging strategies for reducing CCS costs is the development of modular and scalable systems. Modular designs allow components to be prefabricated off-site and assembled quickly, reducing construction time and labor costs. These systems are particularly well-suited for small and medium-scale emitters who cannot afford the high capital outlay of traditional CCS installations.

Scalable solutions can be deployed incrementally, allowing operators to manage financial risk while expanding capacity over time. This approach is increasingly important in developing countries or regions with limited capital markets. Standardized modules also enable better quality control, simplified maintenance, and easier upgrades as technology evolves.

Policy, Financing, and Market-Based Mechanisms

In addition to technological innovations, policy and financial mechanisms play a crucial role in cost reduction. Government subsidies, tax incentives, carbon credits, and low-interest loans can help offset the initial capital investments and make CCS projects more attractive to private investors. For example, the U.S. 45Q tax credit provides financial support for every metric ton of CO_2 captured and stored, thereby improving project economics.

Carbon pricing mechanisms, such as emissions trading systems or carbon taxes, create a financial incentive for emitters to invest in CCS. By assigning a cost to carbon emissions, these policies make low-carbon technologies more competitive and promote innovation. Contracts for difference, where the government guarantees a minimum price for captured CO_2, can also provide revenue certainty and reduce investment risk.

Public-private partnerships and international collaborations are instrumental in spreading risk, pooling resources, and sharing knowledge. Pilot projects funded by governments or development banks can demonstrate feasibility, validate cost estimates, and accelerate the learning curve. As more projects are deployed, the accumulation of operational data and experience will further drive down costs through improved performance and greater confidence in investment decisions.

Lifecycle Cost Analysis and Risk Management

Adopting a lifecycle perspective in project planning and evaluation allows stakeholders to better understand the total cost of CCS projects and identify areas for improvement. Lifecycle cost analysis considers capital investment, operational expenses, maintenance, decommissioning, and long-term monitoring. This holistic approach highlights the importance of early planning, robust design, and proactive risk management in controlling costs. Risk management strategies such as insurance coverage, performance guarantees, and contingency funds are essential for mitigating unexpected expenses. By investing in redundancy, quality control, and real-time monitoring, operators can prevent costly failures and ensure system reliability. Additionally, clear regulatory frameworks and long-term liability arrangements provide investors and operators with the confidence to commit capital to CCS projects.

Public Engagement and Social License to Operate

While not directly related to technological innovation, gaining public support is crucial for reducing delays and legal costs associated with CCS projects. Engaging communities early, maintaining transparency, and addressing concerns about safety, environmental impact, and long-term stewardship can build trust and secure a social license to operate. Projects with strong community support are less likely to face costly opposition or legal challenges, thereby lowering overall project risk and cost.

5.2 FINANCING CCS PROJECTS

5.2.1 Public and Private Funding for CCS R&D

Public and private funding are both essential to advancing CCS R&D and enabling its widespread deployment. While public funding lays the foundation by supporting fundamental research and de-risking early-stage technologies, private investment brings the scale, speed, and innovation needed to move CCS toward commercialization. Collaborative models, policy support, and a shared commitment to climate goals will determine the success of CCS as a viable solution for carbon management. As global emissions targets become more ambitious, expanding and aligning funding for CCS R&D will be crucial in shaping a low-carbon future.

The Role of Public Funding in CCS R&D

Public funding has historically been the cornerstone of CCS R&D, especially during the early stages of technological development when market incentives are limited. Governments recognize that CCS has a vital role to play in achieving national and international climate goals, particularly in sectors where emissions are difficult to eliminate. As a result, public agencies and research institutions are often the first to invest in CCS technologies, supporting fundamental research, pilot projects, and demonstration plants. These investments aim to de-risk emerging technologies, develop technical expertise, and lay the groundwork for commercialization.

Government support comes through various channels, including direct grants, research programs, tax incentives, and public-private partnerships. In countries like the United States, the Department of Energy (DOE) has provided substantial funding through its Office of Fossil Energy and Carbon Management. Initiatives such as the CarbonSAFE program and the National Energy Technology Laboratory's (NETL) carbon storage projects have advanced both capture and storage technologies. Similarly, the European Union has supported CCS R&D through its Horizon Europe program and earlier initiatives like the European Energy Programme for Recovery (EEPR). These programs aim not only to support innovation but also to build infrastructure and regulatory frameworks necessary for deployment.

Public funding often focuses on high-risk, high-reward research that may not attract private investment. It also supports collaboration between universities, national laboratories, and industrial partners, fostering knowledge exchange and capacity building. Furthermore, public investment can guide technological priorities and align R&D efforts with broader societal and environmental objectives. For example, many government-funded CCS projects now emphasize environmental monitoring, stakeholder engagement, and integration with renewable energy sources.

Private Sector Investment in CCS Innovation

Private sector participation in CCS R&D is increasingly important, especially as technologies mature and move closer to commercialization. Large energy companies, industrial firms, and technology developers are investing in CCS to reduce their carbon footprint, comply with regulations, and explore new business models. These companies bring financial resources, technical expertise, and operational experience to the table, accelerating the development and scaling of CCS technologies.

Private funding typically focuses on applied research and pilot projects with clear commercial potential. Companies are particularly interested in improving capture efficiency, reducing energy consumption, and lowering capital and operational costs. Some firms are also exploring ways to monetize captured CO_2, such as through enhanced oil recovery (EOR), the production of synthetic fuels, or the manufacture of carbon-based materials. These commercial applications can provide a revenue stream that offsets R&D costs and incentivizes further investment.

Several multinational corporations, including Shell, Chevron, ExxonMobil, and TotalEnergies, have launched CCS initiatives or invested in CCS startups. These companies often collaborate with universities, research organizations, and government agencies, leveraging public funding to extend the impact of their investments. In some cases, consortia of private and public actors have been formed to develop large-scale CCS hubs that combine capture from multiple sources with shared transportation and storage infrastructure.

Venture capital and private equity have also started to flow into the CCS space, particularly into startups focused on next-generation capture technologies and digital monitoring solutions. Although these investments are still relatively small compared to those in other clean technologies like solar and wind, they reflect growing interest in the commercial potential of CCS. Financial institutions and insurers are also beginning to evaluate the risks and returns of CCS projects, which could open new avenues for debt and equity financing.

Public-Private Partnerships and Collaborative Models

One of the most effective mechanisms for funding CCS R&D is the public-private partnership (PPP) model. These partnerships combine the strengths of both sectors—public funding and oversight with private innovation and execution. PPPs can share risks, align incentives, and ensure that R&D efforts are both scientifically rigorous and commercially relevant. They also promote long-term collaboration and coordination across the value chain, from capture to transport to storage.

Examples of successful PPPs in CCS include the Alberta Carbon Trunk Line in Canada, which connects multiple industrial sources to a common storage site, and the Petra Nova project in Texas, which demonstrated large-scale post-combustion capture on a coal-fired power plant. These projects benefited from a mix of government grants, loan guarantees, tax credits, and private capital. They also provided valuable data and lessons that inform future investments and policy decisions.

International collaborations also play a significant role in CCS R&D. Organizations such as the Carbon Sequestration Leadership Forum (CSLF), the Clean Energy Ministerial, and Mission Innovation bring together governments, companies, and researchers to share knowledge, coordinate funding, and accelerate global deployment. These platforms help avoid duplication of efforts and ensure that advances in one region can benefit others. Multilateral development banks, such as the World Bank and the Asian Development Bank, also provide funding and technical assistance for CCS projects in developing countries.

Challenges in Funding CCS R&D

Despite growing interest, funding for CCS R&D still faces several challenges. One major issue is the long timeline and uncertain return on investment, which makes private investors cautious. Many CCS technologies require significant capital outlay and have long payback periods, which can be a deterrent in competitive markets. Regulatory uncertainty and the absence of strong carbon pricing mechanisms further complicate investment decisions.

Public funding is also subject to political cycles and budget constraints. Shifting policy priorities or changes in government can disrupt funding programs and undermine long-term R&D efforts. In addition, coordination between different agencies and levels of government can be fragmented, leading to inefficiencies and gaps in support. Ensuring consistent and sustained funding across the technology lifecycle is essential for building momentum and achieving deployment at scale.

Another challenge is balancing support for established technologies with investments in novel approaches. While it is important to improve current systems, disruptive innovations may require different funding models and higher risk tolerance. Encouraging a diverse portfolio of R&D activities helps hedge against technological uncertainties and ensures that CCS can adapt to changing energy and environmental conditions.

Future Outlook for CCS Funding

Looking ahead, both public and private funding for CCS R&D are expected to grow as the urgency of climate action increases and the need for deep decarbonization becomes more pressing. National climate strategies, such as the United States' Inflation Reduction Act and the European Green Deal, explicitly support CCS as part of their emissions reduction targets. These policies are likely to generate new funding streams, create market signals, and encourage further investment.

Technological breakthroughs, such as modular capture systems, carbon mineralization, and bioenergy with CCS (BECCS), are expanding the scope and potential of CCS. Continued funding is necessary to bring these innovations from lab to market and to integrate them into broader energy systems. The development of CO_2 transport and storage hubs, particularly in industrial clusters, also presents new opportunities for investment and collaboration.

Increased emphasis on climate finance, sustainability, and ESG (Environmental, Social, and Governance) criteria is prompting investors and corporations to consider CCS in their portfolios. Green bonds, climate funds, and transition finance instruments can provide additional capital for R&D and project development. Engaging stakeholders across the public and private spectrum, including communities, regulators, and civil society, will be critical to ensuring that CCS is deployed in a responsible and equitable manner.

5.2.2 Carbon Pricing and Emission Trading Systems

Carbon pricing and emission trading systems have emerged as essential policy tools in the global effort to reduce greenhouse gas emissions and combat climate change. These mechanisms aim to internalize the social and environmental costs of carbon dioxide emissions by placing a financial cost on emitting activities. The idea is grounded in economic theory, particularly the principle that market-based instruments can effectively correct market failures caused by externalities. By assigning a price to carbon, governments and regulatory bodies create incentives for businesses and individuals to reduce emissions, invest in cleaner technologies, and adopt more sustainable practices.

Carbon pricing can take various forms, including carbon taxes and cap-and-trade systems, each with unique advantages and challenges. Carbon pricing and emissions trading systems represent powerful tools for addressing climate change by aligning economic incentives with environmental goals. By placing a value on carbon emissions, these mechanisms drive innovation, efficiency, and behavioral change, contributing to the transition toward a low-carbon economy. While challenges remain in design, implementation, and public acceptance, the growing adoption and refinement of carbon pricing policies offer a promising path forward. As climate urgency intensifies, expanding and strengthening carbon pricing frameworks will be essential to achieving global emissions targets and fostering sustainable economic development.

Carbon Pricing Mechanisms

Carbon pricing essentially involves charging emitters a fee for every ton of carbon dioxide they emit into the atmosphere. The two primary forms of carbon pricing are carbon taxes and emissions trading systems (ETS), also known as cap-and-trade systems. A carbon tax sets a fixed price per ton of CO_2 emissions, providing certainty about the cost of emissions but leaving the quantity of emissions reductions uncertain. This method is relatively simple to implement and offers predictability in costs for businesses, which can help with planning and investment decisions. In contrast, an emissions trading system sets a cap on the total amount of emissions allowed within a given jurisdiction and allocates or auctions emission permits to companies. These permits can be traded in a marketplace, allowing companies that can reduce emissions at lower costs to sell excess permits to others. This creates a financial incentive for companies to cut emissions and introduces flexibility by enabling the market to find the most cost-effective reduction opportunities. The cap ensures that total emissions stay within environmental targets, providing more certainty in achieving emissions goals than a tax alone.

Emission Trading Systems in Practice

Emission trading systems have been adopted in various forms around the world, most notably in the European Union, California, South Korea, and several regions in China. The European Union Emissions Trading System (EU ETS) is the largest and most established carbon market in the world. It covers major industrial sectors and power generation, requiring companies to hold permits equivalent to their emissions. Over the years, the EU ETS has undergone several phases of reform to improve its effectiveness, including tighter caps, improved allocation methods, and the introduction of a Market Stability Reserve to manage supply and price fluctuations. California's Cap-and-Trade Program, linked with the Canadian province of Quebec, represents a successful subnational carbon market. It covers a wide range of sectors and includes mechanisms to ensure price stability, such as auction price floors and ceilings. The system has demonstrated that emissions trading can be compatible with strong economic growth and can drive significant emissions reductions. China has also launched a national ETS, building on pilot programs in several provinces. Although initially limited to the power sector, China's system is expected to expand over time and play a critical role in meeting its climate commitments.

Carbon Taxes around the World

Carbon taxes are another widely used form of carbon pricing. Countries such as Sweden, Canada, and Switzerland have implemented carbon taxes that have proven effective in reducing emissions while maintaining economic competitiveness.

Sweden's carbon tax, introduced in the early 1990s, is one of the highest in the world and has contributed to significant emissions reductions alongside robust economic growth. Canada has adopted a federal carbon pricing framework that sets a minimum national price on carbon, allowing provinces and territories to implement their own systems or adopt the federal backstop. Carbon taxes provide a clear and direct price signal to polluters, making them easy to understand and administer. They also generate government revenue that can be used to fund clean energy projects, reduce other taxes, or provide rebates to households and businesses. However, setting the right tax level to achieve emissions targets can be challenging, and the effectiveness of a carbon tax depends on how broadly it is applied and how revenues are used.

Economic and Environmental Impacts

The primary goal of carbon pricing is to reduce greenhouse gas emissions by making it more expensive to emit carbon dioxide. By altering the cost structure of fossil fuel-based activities, carbon pricing encourages investment in energy efficiency, renewable energy, and low-carbon technologies. It also promotes behavioral changes among consumers, such as using public transport, reducing energy consumption, or choosing sustainable products. Over time, these shifts can drive substantial emissions reductions across the economy. From an economic perspective, carbon pricing is considered one of the most cost-effective ways to address climate change.

It leverages market mechanisms to allocate resources efficiently and minimizes the overall cost of reducing emissions. When implemented properly, carbon pricing can stimulate innovation, create new industries, and generate government revenue without significantly harming economic growth. Studies have shown that carbon pricing policies can coexist with strong economic performance, particularly when revenues are recycled to support low-income households, reduce other taxes, or invest in green infrastructure.

Challenges and Considerations

Despite its theoretical appeal, carbon pricing faces several practical challenges. One of the main concerns is the potential for adverse economic impacts, particularly on energy-intensive and trade-exposed industries. These industries may face competitive disadvantages if domestic carbon prices are higher than those in other countries, leading to carbon leakage where emissions shift rather than decrease. Policymakers can address this risk through measures such as free allocation of permits, border carbon adjustments, or international coordination of pricing mechanisms.

Equity is another important consideration in carbon pricing. Since energy costs represent a larger share of income for low-income households, carbon pricing can be regressive if not designed carefully. Governments can mitigate these impacts by returning revenues through targeted rebates, tax cuts, or public investments in energy efficiency and public transportation. Transparent and inclusive policy design is essential to building public support and ensuring that carbon pricing contributes to a just transition.

Political feasibility is also a major hurdle. Carbon pricing often faces opposition from interest groups, concerns about rising energy costs, and misconceptions about its impact. Building public awareness, demonstrating the benefits of carbon pricing, and ensuring transparency in how revenues are used are critical to gaining public acceptance. Integrating carbon pricing into broader climate and energy policy frameworks can also enhance its effectiveness and resilience.

Role of Carbon Pricing in Global Climate Policy

As countries seek to meet the goals of the Paris Agreement and transition to net-zero emissions, carbon pricing is increasingly recognized as a key tool in the climate policy toolbox. It provides a flexible, market-based approach that can be scaled and adapted to different national contexts. While not a silver bullet, carbon pricing complements other policies such as regulatory standards, subsidies for clean energy, and research and development support.

Internationally, there is growing interest in coordinating carbon pricing efforts to reduce competitiveness concerns and facilitate global emissions reductions. Initiatives such as the Carbon Pricing Leadership Coalition and the Partnership for Market Readiness support knowledge-sharing and capacity-building among countries. The idea of an international carbon price floor, championed by organizations like the International Monetary Fund, reflects the need for greater alignment and cooperation in carbon pricing policies.

Carbon pricing also plays a role in the development of Article 6 of the Paris Agreement, which outlines rules for international cooperation through carbon markets. Well-functioning international carbon markets can enable countries to achieve emissions reductions more cost-effectively and mobilize finance for climate action in developing countries. However, robust governance, transparency, and environmental integrity are essential to ensure that these mechanisms deliver real and additional emissions reductions.

5.2.3 Carbon Capture Tax Credits and Incentives

Despite their promise, carbon capture tax credits and incentives are not without challenges. One concern is ensuring that public funds are used effectively and that supported projects deliver real, measurable emissions reductions. Rigorous monitoring, reporting, and verification (MRV) systems are needed to track the performance of CCUS projects and ensure compliance. Another issue is the potential for incentives to prolong the life of fossil fuel infrastructure, especially when linked to EOR or other utilization pathways that involve continued fossil fuel production.

To maximize impact, tax credits and incentives should be integrated into a broader policy framework that includes carbon pricing, emissions standards, clean energy mandates, and public investment in infrastructure. Coordinated policies can help avoid unintended consequences, ensure fairness, and align short-term actions with long-term climate goals. Looking ahead, continued innovation in policy design, technology, and business models will be essential to scaling up CCUS. Governments must remain responsive to evolving market conditions, stakeholder needs, and scientific knowledge.

As countries update their nationally determined contributions under the Paris Agreement and pursue deep decarbonization pathways, carbon capture tax credits and incentives will remain a critical part of the policy toolkit. Through targeted support and strategic investment, these measures can help unlock the full potential of carbon capture technologies and accelerate the global transition to a low-carbon future.

Rationale behind Tax Incentives for Carbon Capture

The rationale for offering tax credits and incentives for carbon capture is grounded in the principle of addressing market failures related to climate change. Carbon emissions impose a cost on society through environmental degradation, health impacts, and economic disruption, but these externalities are not reflected in market prices. As a result, there is little economic motivation for industries to reduce emissions voluntarily. Tax incentives and credits help internalize these costs by making carbon capture technologies more financially attractive.

They bridge the gap between the societal benefits of carbon reduction and the private costs borne by project developers, thus encouraging greater deployment of CCUS projects. In addition to addressing market failures, tax credits and incentives also support innovation and technological advancement. The development of carbon capture technologies often requires substantial upfront investments in research, demonstration, and infrastructure. Without government support, many promising technologies may never progress beyond the pilot stage. By lowering financial barriers and providing long-term policy certainty, tax incentives create a more favorable environment for companies to invest in carbon capture solutions, test new approaches, and scale up successful models.

Role of Section 45Q in the United States

One of the most prominent examples of a carbon capture tax incentive is Section 45Q of the U.S. Internal Revenue Code. Introduced in 2008 and significantly expanded in subsequent years, including through the Inflation Reduction Act of 2022, Section 45Q provides a per-ton tax credit for CO_2 that is captured and either permanently stored in geological formations or utilized in qualified applications. The credit is available to both power plants and industrial facilities that install carbon capture equipment and meet certain capture thresholds. Under the revised Section 45Q, the value of the tax credit depends on the end use of the captured CO_2. Projects that sequester CO_2 in secure geological formations, such as saline aquifers or depleted oil fields, receive a higher credit per ton than those that utilize CO_2 for enhanced oil recovery (EOR) or other commercial purposes. The Inflation Reduction Act increased the value of the credit significantly, up to $85 per ton for geological storage and $60 per ton for utilization, making it one of the most generous incentives globally. Furthermore, the Act introduced direct pay and transferability provisions, allowing developers to receive cash payments or transfer the credit to other entities, enhancing liquidity and broadening access.

The improvements to Section 45Q have sparked renewed interest in carbon capture projects across the United States. Power generation, cement, steel, and hydrogen production sectors are now actively exploring CCUS solutions as part of their decarbonization strategies. The predictability and scale of 45Q provide a foundation for long-term planning, which is essential for infrastructure-heavy projects that require significant capital outlay and extended development timelines.

International Incentive Programs

Outside the United States, several countries have implemented tax credits and financial support mechanisms to encourage carbon capture deployment. In Canada, the federal government introduced an investment tax credit (ITC) for CCUS in 2022, offering refundable credits of up to 60 percent for eligible capture equipment and storage expenditures. The Canadian program also emphasizes permanent storage in geological formations and excludes EOR from eligibility, aligning with the country's climate goals. Provinces such as Alberta complement federal efforts with additional funding and regulatory support through programs like the Alberta Carbon Trunk Line, which provides shared infrastructure for CO_2 transport and storage.

The United Kingdom has taken a different approach by focusing on government-backed contracts and capital grants. The UK's CCUS Infrastructure Fund and the Industrial Decarbonisation Challenge provide public funding to support early-stage projects and develop carbon capture clusters in key industrial regions. While not structured as tax credits, these incentives serve a similar function by reducing financial barriers and de-risking investments. The UK government has also proposed a regulated business model for CO_2 transport and storage, ensuring long-term revenue certainty for infrastructure operators.

Other countries such as Norway, Australia, and the Netherlands have also established public funding mechanisms and tax incentives to promote CCUS deployment. In Norway, the government has taken a leadership role in funding projects like Northern Lights, a cross-border CO_2 storage initiative supported by state-owned Equinor, Shell, and TotalEnergies. This project offers a model for international cooperation and public-private partnerships in advancing CCUS.

Design Considerations for Effective Incentives

The effectiveness of carbon capture tax credits and incentives depends on careful policy design. Key considerations include credit value, eligibility criteria, duration, certainty, and administrative simplicity. To stimulate significant investment, the credit must be sufficient to cover the incremental costs of carbon capture compared to business-as-usual operations. At the same time, eligibility rules should be inclusive enough to cover a broad range of sectors and technologies while ensuring that only projects with robust environmental integrity receive support. Long-term policy certainty is essential for attracting investment. Developers are more likely to commit capital to projects when they have confidence that tax credits will be available over the life of the asset. Clear and consistent regulations, coupled with predictable timelines for credit availability, help build investor confidence and reduce the perceived risk of policy changes. Administrative simplicity also plays a role in the effectiveness of incentives. Complex application processes, unclear guidance, or burdensome reporting requirements can deter participation. Governments should strive to streamline procedures, provide technical assistance, and establish transparent monitoring systems to ensure accountability without creating unnecessary barriers.

Economic and Environmental Impacts

When well-designed, tax credits and incentives for carbon capture can deliver substantial economic and environmental benefits. Economically, they stimulate investment in new infrastructure, create high-skilled jobs, and support the competitiveness of domestic industries in a carbon-constrained world. Carbon capture projects often involve complex engineering, construction, and operational activities that generate employment and drive regional economic development. Environmentally, incentives accelerate the deployment of technologies that directly reduce emissions from some of the hardest-to-abate sectors. By enabling the capture and permanent storage of CO_2 from power plants, cement factories, steel mills, and hydrogen production, these policies make a significant contribution to national and global emissions reduction targets. They also support the development of negative emissions technologies, such as direct air capture and bioenergy with carbon capture and storage, which are increasingly seen as necessary to offset residual emissions and achieve net-zero goals.

5.3 POLICY FRAMEWORKS FOR CCS DEPLOYMENT

5.3.1 International Policies on Carbon Reduction

The growing urgency of climate change has led nations around the world to formulate and implement various policies aimed at reducing carbon emissions. International policies on carbon reduction serve as frameworks that encourage, mandate, or facilitate the reduction of greenhouse gas emissions across national borders. These policies reflect global consensus on the need to mitigate climate change impacts and foster sustainable development. While the approaches differ across countries due to varying economic, social, and technological contexts, the overarching objective remains to limit global warming and stabilize atmospheric carbon dioxide levels. As the world moves closer to critical climate thresholds, the need to strengthen international carbon reduction policies becomes increasingly urgent. Future efforts must focus on enhancing ambition, improving implementation, and fostering greater cooperation. This includes revising NDCs to reflect scientific recommendations, scaling up climate finance, and expanding carbon pricing mechanisms. It also requires addressing issues of climate equity, ensuring that the burden of mitigation and adaptation is shared fairly and that vulnerable countries receive the support they need. International policies on carbon reduction represent a dynamic and evolving landscape shaped by scientific knowledge, political negotiation, and social movements. While progress has been made, the path to a low-carbon future demands sustained commitment, innovation, and solidarity among nations. By working together across borders and sectors, the global community can rise to the challenge of climate change and build a more sustainable, resilient, and equitable world for future generations.

Role of the United Nations Framework Convention on Climate Change

At the core of international carbon reduction efforts is the United Nations Framework Convention on Climate Change, which was established in 1992. This treaty set the stage for global cooperation by recognizing the importance of stabilizing greenhouse gas concentrations in the atmosphere. Although it did not impose binding limits, it created a platform for negotiations, policy development, and the mobilization of climate finance. One of the key achievements of the UNFCCC is the Kyoto Protocol, adopted in 1997, which introduced legally binding emission reduction targets for developed countries. The Kyoto Protocol marked the first concrete step towards quantifiable carbon reduction obligations on an international scale, although it was criticized for excluding developing countries and for the limited scope of participation by major emitters like the United States.

Paris Agreement and Nationally Determined Contributions

The most significant milestone in international climate policy came with the adoption of the Paris Agreement in 2015. Unlike the Kyoto Protocol, the Paris Agreement is inclusive and applies to both developed and developing nations. It sets the global goal of limiting the increase in average global temperature to well below 2 degrees Celsius above pre-industrial levels, with efforts to limit the rise to 1.5 degrees. Central to the Paris Agreement are the nationally determined contributions, which are voluntary commitments made by each country outlining how they plan to reduce emissions and adapt to climate change. These NDCs are revised every five years to reflect increasing ambition and improvements in technology and policy. The Paris Agreement also includes provisions for climate finance, capacity building, and technology transfer, especially for vulnerable and developing nations that lack the resources to implement robust carbon reduction strategies.

Carbon Pricing Mechanisms and International Cooperation

One of the most effective tools for reducing carbon emissions is carbon pricing, which assigns a monetary value to greenhouse gas emissions. This economic signal encourages businesses and individuals to reduce emissions and invest in cleaner technologies. Internationally, carbon pricing mechanisms take various forms, including carbon taxes and emissions trading systems. The European Union Emissions Trading System, for example, is one of the largest and most well-established carbon markets in the world. It sets a cap on total emissions and allows companies to buy and sell allowances, thereby incentivizing reductions where they are most cost-effective. International cooperation on carbon pricing is growing, with discussions underway about linking different national and regional systems to create a global carbon market. Such linkage would increase the efficiency of emissions reductions by enabling broader trading opportunities and harmonizing carbon prices across borders. Moreover, international financial institutions like the World Bank and the International Monetary Fund have been vocal advocates for carbon pricing as a key strategy in global carbon reduction policies.

Sector-Specific Policies and Multilateral Agreements

International carbon reduction policies also include sector-specific initiatives and multilateral agreements. In the aviation sector, the Carbon Offsetting and Reduction Scheme for International Aviation was adopted by the International Civil Aviation Organization to cap net emissions from international flights. CORSIA requires airlines to offset emissions growth beyond 2020 levels by purchasing carbon credits from approved programs. Similarly, the International Maritime Organization has adopted strategies to reduce carbon emissions from shipping through measures such as energy efficiency standards and low-carbon fuel mandates. In the energy sector, global initiatives like the Clean Energy Ministerial and Mission Innovation aim to accelerate the deployment of renewable energy technologies and improve energy efficiency. These efforts are complemented by bilateral and multilateral partnerships that promote knowledge sharing, technical assistance, and joint research projects. Such cooperation is essential in enabling developing countries to leapfrog fossil-fuel-dependent development models and transition to cleaner, more sustainable energy systems.

Adaptation, Resilience, and Climate Finance

While carbon reduction is a priority, international policies also recognize the importance of adaptation and resilience, especially for countries that are most vulnerable to the impacts of climate change. The Paris Agreement includes a strong emphasis on supporting adaptation efforts through climate finance and international cooperation. Developed countries have pledged to mobilize $100 billion per year to assist developing nations in mitigating emissions and building resilience. This financial support is critical for implementing low-carbon development pathways, improving infrastructure, and safeguarding communities against climate-related disasters. In addition to direct funding, international institutions have created climate finance mechanisms such as the Green Climate Fund and the Global Environment Facility. These institutions provide grants, loans, and technical support for climate action projects worldwide. However, challenges remain in terms of ensuring equitable access to these resources and increasing the scale and predictability of funding. Greater transparency, efficiency, and accountability in climate finance governance are needed to build trust among nations and drive more ambitious climate action.

Challenges in Implementation and Compliance

Despite the progress made through international policies, significant challenges persist in implementation and compliance. One major issue is the voluntary nature of the Paris Agreement, which relies on political will and peer pressure rather than binding enforcement mechanisms. This can lead to discrepancies between stated commitments and actual performance, with some countries failing to meet their NDC targets or delaying action. Furthermore, geopolitical tensions, economic instability, and competing national priorities often hinder global consensus and coordinated action. Another challenge is the need for accurate measurement, reporting, and verification of emissions and mitigation efforts. Without robust MRV systems, it is difficult to assess progress, hold countries accountable, and ensure the integrity of carbon reduction claims. Investments in data infrastructure, technical capacity, and transparency mechanisms are therefore essential for enhancing the credibility and effectiveness of international carbon reduction policies.

Role of Non-State Actors and Civil Society

International policies on carbon reduction are no longer confined to national governments. Non-state actors, including cities, businesses, and civil society organizations, play a vital role in advancing climate action. Initiatives like the Race to Zero campaign, the Science Based Targets initiative, and the Powering Past Coal Alliance demonstrate the growing commitment of sub-national and private sector entities to reducing emissions. These actors often act as laboratories for innovation, testing new policies and technologies that can later be adopted at larger scales. Civil society also plays a crucial role in holding governments accountable, raising public awareness, and advocating for stronger climate policies. International climate summits and forums provide opportunities for engagement, collaboration, and knowledge exchange among diverse stakeholders. The inclusion of indigenous communities, youth organizations, and marginalized groups in climate governance helps ensure that policies are inclusive, equitable, and grounded in social justice.

5.3.2 CCS Regulations: Safety, Liability, and Monitoring

The regulatory framework for CCS plays a pivotal role in ensuring the safety, reliability, and public acceptance of carbon storage projects. By addressing concerns related to operational safety, long-term liability, and comprehensive monitoring, regulations provide the foundation for responsible deployment of this critical climate mitigation technology. As the global community continues to confront the challenges of climate change, strengthening and harmonizing CCS regulations will be essential to unlocking the full potential of carbon capture and storage as a viable and scalable solution.

Ensuring Safety in CCS Operations

Safety is paramount in all stages of a CCS project, from the initial capture of CO_2 to its transportation and ultimate storage. Regulatory frameworks around the world place strict safety requirements on the selection of storage sites, construction of infrastructure, and operational practices. Site selection is governed by geological criteria that ensure the presence of impermeable caprock layers, sufficient storage capacity, and minimal risk of seismic activity or leakage. Before a site is approved, detailed geological surveys, modeling, and risk assessments must be conducted to evaluate its suitability and integrity over decades or centuries.

The transportation of CO_2, usually in compressed form via pipelines or ships, is regulated under standards similar to those used in the transport of natural gas or other hazardous materials. Regulations require that pipelines be designed to withstand pressure fluctuations, corrosion, and physical damage. Operational protocols must be in place to manage accidental releases, which could pose health risks or damage ecosystems. In addition, emergency response systems and regular inspections are mandatory to detect potential threats early and respond effectively.

Once injected into storage formations, the CO_2 must remain securely trapped to prevent its re-entry into the atmosphere. Regulators require stringent operational controls, including managing injection pressures to avoid fracturing the caprock or inducing seismicity. The use of advanced engineering practices, such as real-time pressure monitoring and injection rate controls, helps ensure that storage conditions remain within safe operational boundaries. Safety regulations also cover worker training, facility design, and contingency planning, reinforcing a culture of risk management across the CCS lifecycle.

Liability and Long-Term Stewardship

The issue of liability is central to the legal and regulatory architecture of CCS. Unlike other industrial operations, the risks associated with CCS extend far into the future, well beyond the active life of the storage facility. Determining who is responsible if stored CO_2 leaks decades after injection has ceased is a critical concern that influences investor confidence, insurance provisions, and public support. Regulatory regimes must therefore define clear rules around short-term and long-term liability, including who holds financial and legal responsibility for monitoring, remediation, and potential damages.

In most jurisdictions, operators are initially responsible for the integrity and safety of the storage site. They must adhere to strict conditions laid out in their permits and demonstrate compliance through continuous reporting and inspection. After closure of the injection operation, a post-closure period follows during which the operator must continue to monitor the site and manage any residual risks. This period can range from decades to potentially a century, depending on the stability of the storage formation and the regulatory requirements.

Eventually, once the operator can demonstrate that the stored CO_2 is permanently contained and that the site poses no significant risk to the environment or public health, liability may be transferred to the government. This transfer is contingent on meeting regulatory performance standards and providing robust evidence of long-term stability. To prepare for this eventuality, many countries require operators to establish financial assurance mechanisms, such as trust funds, insurance policies, or bonds, to cover the costs of monitoring, remediation, or compensation should issues arise after closure.

The debate over long-term liability is particularly relevant in cross-border CCS projects, where CO_2 captured in one country may be transported and stored in another. In such cases, international agreements and harmonized regulatory standards are needed to allocate responsibilities and avoid legal disputes. The London Protocol and the EU CCS Directive are examples of efforts to address these transboundary challenges, providing legal frameworks that guide the development of secure and accountable CCS operations across multiple jurisdictions.

Comprehensive Monitoring and Verification Requirements

Monitoring is a cornerstone of any CCS regulatory system. It ensures that stored CO_2 remains contained, allows for early detection of potential leaks, and builds public and regulatory confidence in the technology. Regulatory authorities require operators to develop and implement detailed monitoring plans, which must be approved before injection begins. These plans outline the types of monitoring technologies to be used, the frequency of measurements, and the parameters to be tracked over time.

Monitoring involves both surface and subsurface techniques. Subsurface monitoring typically includes seismic imaging, pressure and temperature sensors, and geochemical sampling. These methods allow operators to track the movement of CO_2 within the geological formation and verify that it is behaving as predicted. Surface monitoring may involve soil gas surveys, atmospheric sensors, and groundwater testing to detect any signs of leakage to the environment. Satellite imaging and remote sensing technologies are increasingly being used to provide wide-area surveillance and long-term observational data.

Regulations also require the use of baseline data, which is collected before injection starts to establish reference conditions. This baseline is critical for distinguishing between natural variations and potential impacts from CO_2 injection. Any deviations from expected behavior trigger regulatory responses, which may include additional monitoring, operational adjustments, or corrective actions. In extreme cases, injection operations may be suspended until the issue is resolved.

Verification is the process of independently assessing the monitoring data to ensure compliance with regulatory requirements. Regulatory agencies or third-party auditors review the data and inspect the site to confirm that storage integrity is being maintained. Regular reporting to authorities is mandatory, and in some cases, information must be made publicly available to maintain transparency and accountability.

Evolving Regulatory Frameworks and Global Harmonization

As CCS technology matures and deployment scales up, regulatory frameworks are evolving to keep pace with new challenges and innovations. Initially, most CCS regulations were adapted from existing oil and gas or hazardous waste laws. While this provided a starting point, many countries have since developed dedicated CCS legislation to address the unique aspects of carbon storage. For example, the United States Environmental Protection Agency developed the Class VI well permitting system under the Safe Drinking Water Act specifically for geologic sequestration, requiring rigorous site characterization and monitoring protocols.

In the European Union, the CCS Directive establishes a harmonized legal framework for the safe storage of CO_2 across member states. It mandates detailed requirements for site selection, monitoring, liability transfer, and public participation. Other countries, including Canada, Australia, and Japan, have introduced similar regulations, tailored to their geological conditions and legal systems. These efforts contribute to a growing body of best practices and shared knowledge, facilitating more consistent and effective regulation worldwide.

Global organizations such as the International Energy Agency, the Global CCS Institute, and the Carbon Sequestration Leadership Forum play vital roles in promoting harmonization of standards and regulatory cooperation. Their work helps align national approaches, foster information exchange, and build capacity in countries seeking to develop their own CCS regulations. International collaboration is particularly important for ensuring that CCS can be deployed at the scale and speed needed to meet global climate goals.

Public Trust, Transparency, and Regulatory Integrity

A key factor in the success of CCS is public trust, which is closely tied to the transparency and integrity of regulatory systems. Communities near storage sites must have confidence that their safety and environmental well-being are protected by robust and enforceable regulations. Public engagement is therefore an integral part of regulatory development and project approval processes. Authorities often require public consultations, environmental impact assessments, and stakeholder involvement as part of the permitting procedure. Transparency in monitoring results, regulatory decisions, and incident reporting further reinforces public confidence. Many jurisdictions mandate the publication of monitoring data and inspection reports, enabling independent review and scrutiny. Strong enforcement mechanisms, including penalties for non-compliance and the authority to suspend or revoke permits, underscore the seriousness of regulatory obligations and deter negligent practices.

5.3.3 National vs. Global CCS Policies

National vs. Global CCS Policies

Carbon Capture and Storage (CCS) has emerged as a critical strategy in combating climate change by capturing carbon dioxide (CO_2) emissions from industrial processes and power generation and securely storing them in geological formations. As the world seeks to transition toward a low-carbon future, both national and global policies have taken shape to regulate, promote, and coordinate the implementation of CCS technologies. While national policies are tailored to a country's economic, legal, and environmental contexts, global policies strive to create a unified framework that enables international cooperation and shared technological development. Understanding the dynamics between national and global CCS policies is essential for ensuring that carbon capture and storage initiatives are implemented effectively and contribute meaningfully to international climate goals.

Foundation of National CCS Policies

At the national level, governments formulate CCS policies based on domestic priorities, resource availability, and regulatory capacities. National CCS policies generally involve legal frameworks, financial incentives, permitting procedures, and research and development support. Countries with substantial fossil fuel industries, such as the United States, Canada, Australia, and China, have developed relatively advanced CCS frameworks. These nations recognize the role CCS plays not only in emissions reduction but also in maintaining energy security and economic competitiveness. For instance, the United States has enacted tax credit systems like Section 45Q, which incentivizes CO_2 capture and storage by offering financial rewards per metric ton of CO_2 sequestered. Additionally, the U.S. Environmental Protection Agency regulates CCS under the Underground Injection Control (UIC) Program, ensuring that storage activities protect underground sources of drinking water. Canada has similarly developed robust CCS policies, particularly in provinces like Alberta, where regulations align with the Oil and Gas Conservation Act and include monitoring and liability frameworks to manage long-term storage risks.

National policies also support large-scale demonstration projects to evaluate the feasibility and cost-effectiveness of CCS in real-world conditions. These pilot projects serve as testing grounds for regulatory procedures, technological performance, and stakeholder engagement strategies. Moreover, national governments often collaborate with industry and academia to establish CCS research hubs, innovation networks, and public-private partnerships that drive technological advancement.

Challenges of Disparate National Approaches

Despite progress, the variability in national CCS policies presents several challenges. Different countries operate under diverse legal systems, economic structures, and environmental standards, resulting in inconsistent approaches to CCS regulation and deployment. Some nations emphasize strict regulatory oversight with rigorous environmental assessments, while others prioritize rapid deployment with fewer barriers. This lack of harmonization complicates cross-border CCS initiatives and the development of international carbon markets.

The inconsistency is particularly problematic when considering the transportation and storage of CO_2 across national boundaries. For example, a country capturing CO_2 may lack adequate geological storage capacity and need to export it to another country. In the absence of aligned national policies and legal agreements, such transactions become legally and logistically difficult. Additionally, differences in liability frameworks, site closure protocols, and monitoring standards create uncertainties for investors and technology developers seeking to operate across multiple jurisdictions.

Furthermore, countries with limited financial or technical capacity often struggle to develop and implement CCS policies, leading to a geographic concentration of projects in wealthier or more industrialized nations. This imbalance risks exacerbating global inequities in climate mitigation efforts and hinders the overall scalability of CCS as a global solution. To bridge this gap, international cooperation and funding support are required to help developing nations build the institutional and technical frameworks necessary for CCS deployment.

Global CCS Policies and Agreements

Global CCS policies are primarily shaped through international climate agreements, multilateral organizations, and cooperative initiatives. These policies aim to establish common goals, facilitate technology transfer, and promote coordinated action across nations. The United Nations Framework Convention on Climate Change (UNFCCC) and the Paris Agreement are central platforms that provide overarching targets for greenhouse gas reductions and recognize the potential of CCS in achieving those targets.

Under the Paris Agreement, countries submit Nationally Determined Contributions (NDCs) outlining their individual climate actions. Although the inclusion of CCS in NDCs varies by country, global climate negotiations have acknowledged the importance of CCS in decarbonizing hard-to-abate sectors such as cement, steel, and petrochemicals. Global forums encourage countries to integrate CCS into their mitigation strategies and report progress through standardized frameworks.

International organizations such as the International Energy Agency (IEA), the Global CCS Institute, and the Carbon Sequestration Leadership Forum (CSLF) play critical roles in shaping global CCS policy. These bodies conduct research, facilitate dialogue, and provide technical assistance to member states. Their work supports the development of best practices, harmonized monitoring standards, and capacity-building programs that enable global alignment on CCS deployment.

Multilateral climate finance mechanisms, including the Green Climate Fund and the Climate Investment Funds, offer financial support for CCS projects in developing countries. These mechanisms are vital for ensuring equitable access to CCS technology and expanding its application beyond high-income nations. They also encourage collaboration between developed and developing countries, fostering joint ventures and knowledge-sharing initiatives.

Synergies and Tensions between National and Global Policies

While national and global CCS policies share the common objective of reducing CO_2 emissions, their interaction can lead to both synergies and tensions. On one hand, global frameworks provide guidance, legitimacy, and financial resources that support national implementation. They help align national goals with international climate targets and encourage countries to adopt robust CCS regulations and incentives. Global partnerships also enable the transfer of knowledge, infrastructure, and expertise, accelerating technological development across borders. On the other hand, national sovereignty and political considerations can limit the influence of global CCS policies. Countries may prioritize domestic energy interests, industrial competitiveness, or political ideologies over global climate commitments. This can result in a lack of policy coherence, delayed implementation, or resistance to international oversight. Furthermore, national regulatory systems often evolve independently, making it difficult to achieve uniformity in monitoring, verification, and reporting standards.

Another source of tension arises from differing views on the role of CCS in the broader decarbonization strategy. Some nations view CCS as a transitional technology that supports continued fossil fuel use, while others emphasize renewable energy and energy efficiency as long-term solutions. These divergent perspectives influence national policy choices and affect the level of commitment to global CCS cooperation.

5.4 CCS IN CARBON MARKETS

5.4.1 Role of Carbon Capture in Carbon Trading Markets

The role of carbon capture in carbon trading markets represents a significant intersection of technology and policy aimed at combating climate change. By enabling the issuance of high-quality carbon credits, CCS enhances the effectiveness and credibility of emissions trading systems. It offers a practical solution for industries with limited decarbonization options and supports the broader objective of achieving net-zero emissions. While challenges related to cost, verification, and policy integration remain, ongoing efforts in research, regulation, and international cooperation are paving the way for carbon capture to become a cornerstone of carbon trading and global climate strategy. As the urgency of climate action intensifies, the alignment of carbon capture technologies with market-based mechanisms will be essential to achieving deep, lasting reductions in global greenhouse gas emissions.

Carbon Trading Mechanisms and Market Structures

Carbon trading markets operate under two main systems: cap-and-trade programs and voluntary carbon markets. In a cap-and-trade system, governments set a limit or cap on total greenhouse gas emissions allowed in a specific region or sector. Companies receive or purchase emission allowances, which they can trade with others. If a company emits less than its allowance, it can sell the surplus; if it exceeds the limit, it must buy additional permits or face penalties. This creates a financial incentive for reducing emissions. In contrast, voluntary carbon markets enable businesses and individuals to offset their emissions by purchasing carbon credits generated from emission-reduction projects, including those involving CCS, reforestation, or renewable energy development.

Carbon Capture as a Source of Carbon Credits

Carbon capture technologies can generate tradable carbon credits when they successfully prevent CO_2 from entering the atmosphere. These credits represent verified emissions reductions and can be sold in both compliance and voluntary carbon markets. For CCS projects to issue carbon credits, they must undergo rigorous monitoring, reporting, and verification (MRV) procedures to ensure that the CO_2 is permanently sequestered and not leaked back into the environment. Once verified, these projects contribute to the carbon market by providing a new source of high-quality offsets, helping emitters meet their regulatory or voluntary commitments. This potential to monetize captured carbon incentivizes investment in CCS infrastructure and innovation.

Enhancing Market Value and Credibility

The inclusion of carbon capture in carbon trading markets not only provides additional supply of carbon credits but also enhances the credibility and robustness of the market. CCS is one of the few solutions that can deliver permanent emissions reductions, which is particularly valuable in a market often criticized for relying on temporary or questionable offset projects. By introducing high-integrity credits from CCS, the market gains in environmental integrity, thereby boosting investor confidence and attracting greater participation. Furthermore, CCS credits are particularly suited for long-term climate strategies as they align with the goal of achieving net-zero and even net-negative emissions in the future.

Encouraging Private Sector Participation

Carbon trading markets linked with carbon capture create a pathway for private sector participation in climate action. By establishing a price signal through the trading of carbon credits, these markets encourage companies to adopt CCS as a cost-effective strategy for compliance and corporate sustainability goals. For instance, oil and gas companies with high emissions profiles can invest in CCS to meet emissions targets while continuing operations. Additionally, emerging direct air capture (DAC) technologies that extract CO_2 directly from the atmosphere can generate negative emissions credits, appealing to firms that seek to go beyond net-zero and actively remove their historical carbon footprint.

Challenges in Integration and Standardization

Despite its potential, the integration of carbon capture into carbon trading markets faces several challenges. One of the key issues is the standardization of methodologies for measuring and verifying emissions reductions from CCS projects. Different carbon standards and registries may apply varying criteria, leading to inconsistencies in credit issuance and market acceptance. Moreover, the high upfront costs and technical complexity of CCS projects can deter investment, particularly in regions where carbon prices are low or where regulatory support is insufficient. Legal and liability concerns regarding long-term storage integrity and potential CO_2 leakage also present barriers that need to be addressed through robust legal and monitoring frameworks.

International Cooperation and Policy Support

The role of carbon capture in carbon trading markets is increasingly recognized at the international level. Institutions such as the United Nations Framework Convention on Climate Change (UNFCCC) have incorporated CCS in their mechanisms under Article 6 of the Paris Agreement, which enables international cooperation through carbon credit trading. Multilateral support from organizations like the World Bank and the International Energy Agency (IEA) further promotes CCS deployment by providing technical assistance, financial tools, and capacity building. Governments are also introducing policies such as tax credits, grants, and public-private partnerships to stimulate investment in carbon capture projects and integrate them into national emissions trading schemes.

Potential for Market Expansion and Innovation

As global demand for credible and scalable carbon removal solutions grows, carbon capture is poised to play a central role in the expansion of carbon trading markets. New market mechanisms are emerging that specifically cater to carbon removal, distinguishing it from emissions avoidance or reduction. This includes the development of premium credit markets that value the permanence and additionality of technologies like CCS and DAC. Additionally, innovation in blockchain and digital MRV tools can streamline the verification process, reduce transaction costs, and increase transparency in carbon credit trading, thereby accelerating market adoption.

5.4.2 Carbon Offsets and Verifying CCS Credits

Carbon offsets have become an essential component in the global strategy to combat climate change, offering businesses, governments, and individuals a mechanism to compensate for greenhouse gas emissions by supporting projects that reduce or remove emissions elsewhere. One of the most advanced and increasingly important sources of carbon offsets is carbon capture and storage, or CCS. As climate policies and carbon markets evolve, the credibility and transparency of carbon offsets generated through CCS are vital. This has led to a strong emphasis on verification mechanisms that can ensure the legitimacy and permanence of CCS credits. Carbon offsets are a critical tool in the global response to climate change, and carbon capture and storage has emerged as a high-integrity source of such offsets. However, the environmental effectiveness and market credibility of CCS-based carbon credits depend heavily on rigorous verification processes. Through comprehensive monitoring, transparent reporting, and independent third-party validation, the legitimacy of CCS credits can be assured. While challenges remain in terms of technical complexity and policy alignment, the future of verified CCS credits is promising. As carbon markets continue to expand and evolve, the integration of verified CCS offsets will be essential to achieving deep and lasting reductions in atmospheric carbon dioxide and delivering on global climate goals.

Carbon Offsets and Their Purpose

Carbon offsets function as compensatory mechanisms that allow entities to balance out their greenhouse gas emissions by investing in projects that reduce or remove an equivalent amount of emissions from the atmosphere. These projects can include renewable energy installations, afforestation or reforestation efforts, methane capture, energy efficiency improvements, and more recently, carbon capture and storage initiatives. Offsets are measured in metric tons of carbon dioxide equivalent and are traded in both compliance markets and voluntary markets. In compliance markets, organizations are legally required to offset a portion of their emissions under emission trading schemes. In voluntary markets, businesses or individuals choose to offset their carbon footprint to meet corporate social responsibility goals or sustainability commitments.

CCS in Generating Carbon Offsets

Carbon capture and storage plays a unique role in the generation of carbon offsets due to its ability to directly capture CO_2 from industrial processes or the atmosphere and store it permanently in geological formations. This distinguishes CCS-based offsets from many other types of projects which often focus on emissions avoidance rather than actual removal. The process typically involves capturing CO_2 at the source, transporting it via pipelines or ships, and injecting it into suitable underground storage sites. When properly executed, this process results in permanent sequestration of CO_2, thereby qualifying it as a reliable source of carbon credits. Especially in sectors like cement, steel, and power generation, where emissions are otherwise difficult to eliminate, CCS provides a realistic pathway to meaningful emissions reduction.

Importance of Verifying CCS Credits

Verification is the cornerstone of trust in the carbon market. Without robust verification, the environmental integrity of carbon offsets would be in question. For CCS credits, verification is especially critical because the process involves highly technical operations and long-term storage commitments. Verification ensures that the CO_2 claimed to be captured has indeed been removed from the atmosphere and stored securely. It also certifies that there are no significant leakages or reversals of the stored CO_2 over time. This is achieved through the use of monitoring, reporting, and verification (MRV) protocols. Effective MRV frameworks are designed to assess the performance of CCS projects, confirm that operational standards are met, and track the fate of the captured CO_2 over extended periods.

Monitoring and Reporting Requirements

Monitoring in CCS projects involves continuous observation of the entire CO_2 capture, transport, and storage process. Operators must collect detailed data on the volume of CO_2 captured, the composition of the gas stream, transport dynamics, and the conditions of injection at storage sites. Once injected, the storage site must be closely monitored using seismic surveys, well logging, pressure and temperature sensors, and chemical tracers to ensure the CO_2 remains safely contained. This information is compiled into periodic reports submitted to regulatory authorities or independent verifiers. Transparent and accurate reporting is essential not only for credit issuance but also for maintaining public and market confidence in the environmental benefits of CCS.

Third-Party Verification and Certification Standards

To ensure impartiality and consistency, third-party verification bodies are employed to validate CCS project performance and the corresponding carbon credits. These verifiers assess project documentation, audit data collection processes, evaluate monitoring techniques, and inspect operational compliance. Leading international standards for carbon credits, such as the Verified Carbon Standard (VCS), the Gold Standard, and the American Carbon Registry (ACR), have developed methodologies specific to CCS. These methodologies outline the eligibility criteria, baseline setting, additionality requirements, and permanence considerations for projects to qualify for offset generation. By adhering to these certification standards, CCS projects can issue high-quality, tradable carbon credits that are widely accepted in both compliance and voluntary carbon markets.

Ensuring Additionality and Permanence

Two of the most important principles in the verification of carbon offsets are additionality and permanence. Additionality means that the emissions reductions would not have occurred without the offset project. For CCS, this implies that the carbon capture and storage activity must go beyond business-as-usual practices and must be implemented specifically to reduce emissions in a measurable way. Permanence refers to the long-term reliability of the emissions reduction. In the case of CCS, permanence is ensured by selecting suitable geological formations, maintaining rigorous operational protocols, and implementing long-term monitoring programs. In addition, many standards require the establishment of risk mitigation measures, such as buffer credit reserves or liability funds, to compensate for any future reversals or leakages.

Challenges in Verifying CCS Credits

Despite the sophistication of CCS technologies and verification methodologies, several challenges persist. The complexity of geological storage and the need for continuous monitoring over decades create logistical and financial burdens. Establishing baseline emissions for industrial processes can be difficult, particularly when operations are variable or not fully transparent. Another challenge is demonstrating additionality in a policy landscape that is rapidly evolving to mandate emissions reductions. There is also the issue of public perception and social license, as communities may harbor concerns about underground CO_2 storage and the potential for leaks. Addressing these concerns requires transparent communication, stakeholder engagement, and continuous improvements in verification practices.

Regulatory and Policy Support for Verification

Effective verification of CCS credits is heavily dependent on supportive regulatory frameworks. National governments and international bodies play a crucial role in defining legal standards for CO_2 storage, setting liability rules, and mandating MRV requirements. In jurisdictions with emissions trading schemes, such as the European Union or California, regulators have developed detailed protocols for the inclusion of CCS in compliance markets. Moreover, international cooperation under agreements like the Paris Agreement encourages harmonization of MRV standards and supports the development of cross-border carbon credit trading mechanisms. Policy incentives, such as tax credits and subsidies, further enhance the economic feasibility of verified CCS projects.

5.5 PUBLIC PERCEPTION AND STAKEHOLDER ENGAGEMENT

5.5.1 Communicating the Importance of CCS

Communicating the importance of CCS is not merely a technical challenge; it is a strategic imperative for climate action. As governments, industries, and communities seek pathways to net-zero emissions, understanding and accepting CCS will be critical to mobilizing the necessary support and investment. Effective communication requires clarity, transparency, empathy, and adaptability. It involves translating complex science into everyday language, addressing public concerns with honesty, and inspiring confidence through real-world successes. As the climate crisis deepens and the urgency for solutions grows, ensuring that CCS is widely understood and supported will be essential to achieving the emissions reductions needed to safeguard the planet for future generations.

Public Awareness in Climate Action

Public awareness and understanding play a pivotal role in shaping attitudes toward climate solutions. In the context of CCS, the lack of widespread knowledge or interest can lead to skepticism, opposition, or underestimation of its relevance. People tend to support climate solutions that they understand and perceive as beneficial, safe, and aligned with their values. Hence, one of the first steps in communicating the importance of CCS is to raise awareness about its role in the broader climate strategy. This means placing CCS in the same category as renewable energy, energy efficiency, and sustainable agriculture, highlighting that it is not a standalone solution but part of an integrated effort to achieve net-zero emissions. It is also vital to correct misconceptions that CCS is only a tool to extend the life of fossil fuel industries, when in fact, it is essential for decarbonizing sectors with few other viable alternatives.

Technology in Accessible Language

Communicating complex scientific concepts like CCS requires translation into accessible language that resonates with a wide range of audiences. The technical steps of capturing carbon dioxide at its source, transporting it to a storage site, and injecting it deep underground for permanent storage must be described in ways that non-experts can grasp. Effective communication avoids jargon and instead uses analogies, real-world examples, and visual storytelling to convey how CCS works. For example, explaining that storing CO_2 underground is similar to how natural gas has been stored safely for decades helps build familiarity and reduce perceived risks. Infographics, videos, and interactive tools can be powerful resources to engage different segments of the public, from school children to policymakers.

Building Trust through Transparency and Dialogue

Trust is a critical factor in public acceptance of any large-scale infrastructure project, especially those involving environmental and safety concerns. In the case of CCS, fears about potential CO_2 leaks, earthquakes, or other unintended consequences can fuel opposition if not addressed proactively. Transparent communication, based on facts and open dialogue, is key to building trust. Project developers, governments, and scientists must be willing to engage with communities, listen to concerns, and provide honest answers supported by data and experience. Trust is also built through consistent monitoring, third-party verification, and public reporting of CCS operations. Demonstrating a track record of safe and effective CCS implementation in different parts of the world can help overcome hesitation and foster greater acceptance.

Highlighting the Urgency and Necessity of CCS

A central message in communicating CCS is its critical importance in meeting global climate goals. According to leading scientific bodies such as the Intergovernmental Panel on Climate Change (IPCC) and the International Energy Agency (IEA), achieving net-zero emissions by mid-century will be extremely difficult, if not impossible, without large-scale CCS deployment. This is particularly true for industries where emissions cannot be entirely eliminated through renewable energy or electrification. Communicators should emphasize that CCS is not a substitute for renewable energy but a complementary tool that enables emissions reductions across a broader range of sectors. It also has the unique potential to deliver negative emissions when combined with bioenergy or direct air capture, making it essential not just for slowing emissions but for removing carbon already in the atmosphere.

Tailoring Messages for Different Audiences

Effective communication about CCS requires tailored approaches that consider the interests, values, and knowledge levels of different audiences. For policymakers, the emphasis may be on regulatory frameworks, economic competitiveness, and job creation. For investors, messages might focus on financial returns, risk management, and long-term value. For local communities, the key concerns often revolve around environmental safety, health impacts, and benefits such as employment or infrastructure development. Academic audiences may be more interested in technical innovation and research opportunities, while youth and environmental activists might be more responsive to arguments about climate justice and intergenerational responsibility. Understanding the audience and framing messages accordingly is crucial to ensuring that communication about CCS is both relevant and persuasive.

Leveraging Media and Digital Platforms

The media, both traditional and digital, plays a significant role in shaping public perceptions of CCS. Journalists, influencers, and content creators act as intermediaries who can either amplify support for CCS or spread misinformation. Building relationships with media professionals and providing them with accurate, engaging content is essential for reaching broader audiences. At the same time, social media platforms offer powerful tools for outreach and education. Organizations involved in CCS can use platforms like Twitter, YouTube, and LinkedIn to share project updates, publish explainer videos, host live Q&A sessions, and connect with followers. A consistent and engaging online presence can help maintain interest in CCS and counteract negative narratives or misinformation.

Integrating CCS in Climate Education and Campaigns

Long-term support for CCS depends on integrating it into broader climate education and advocacy. Schools, universities, and environmental organizations have a role to play in incorporating CCS into curricula, workshops, and public campaigns. By including CCS in climate change discussions from an early age, future generations will have a more comprehensive understanding of the tools available to address the crisis. Climate advocacy campaigns that traditionally focus on renewable energy, conservation, and policy reform should also include messaging about the importance of CCS to create a more balanced and informed narrative. This integration can also strengthen alliances across different sectors of the climate movement, uniting them around the shared goal of deep decarbonization.

Communicating Real-World Success Stories

Highlighting successful CCS projects can be one of the most effective ways to demonstrate the technology's value and feasibility. Real-world examples, such as Norway's Sleipner project, Canada's Boundary Dam facility, or the Petra Nova plant in the United States, provide tangible proof that CCS works and can be deployed safely. These stories should include data on emissions captured, community engagement efforts, economic impacts, and lessons learned. Communicators should also spotlight innovative projects that combine CCS with renewable energy, hydrogen production, or bioenergy, showcasing the technology's versatility. By focusing on real outcomes and human experiences, these narratives can inspire confidence and support for future projects.

Overcoming Misconceptions and Opposition

CCS has faced criticism from some quarters of the environmental movement, often based on concerns that it prolongs the use of fossil fuels or diverts attention and resources from renewable energy. Addressing these concerns requires acknowledging the valid critiques while clarifying the unique role CCS plays. Communicators should emphasize that CCS is essential for sectors where other solutions are not viable and that its use is fully compatible with a broader transition to clean energy. In fact, the deployment of CCS can accelerate emissions reductions while supporting economic stability and job preservation during the energy transition. Clear, fact-based responses to opposition can help shift the conversation from ideological conflict to pragmatic cooperation.

5.5.2 Addressing Public Concerns about Safety and Effectiveness

Addressing public concerns about the safety and effectiveness of CCS is not simply about defending a technology; it is about fostering trust, inclusivity, and shared responsibility in the fight against climate change. By grounding communications in science, engaging communities openly, ensuring robust legal and monitoring frameworks, and integrating ethical considerations, stakeholders can create a foundation for public confidence in CCS. As the urgency to reduce emissions grows, building societal support for technologies like CCS will be essential. The success of CCS depends not only on engineering but also on earning and maintaining the trust of the public it aims to serve.

Public Concerns

Public apprehension regarding CCS is driven by both technical uncertainties and broader socio-political factors. Many people are unfamiliar with the mechanics of carbon capture, transport, and underground storage, and thus may view the entire process as experimental or risky. Fears of CO_2 leakage from storage sites, accidental releases during transportation, and potential seismic activity due to underground injection can lead to resistance, especially in areas near proposed project sites. Moreover, the historical mistrust of industrial projects, especially those involving underground operations like fracking or mining, contributes to heightened vigilance and community opposition. Concerns also emerge from the perception that CCS may serve as a means to prolong fossil fuel dependence under the guise of climate action, thereby diverting attention from cleaner energy alternatives. Understanding these concerns in their full complexity is the first step toward effectively addressing them.

Scientific Basis of CCS Safety

To allay public fears, it is essential to communicate the strong scientific foundation that underpins CCS safety protocols. The geological formations used for CO_2 storage, such as depleted oil and gas reservoirs or deep saline aquifers, have been studied extensively and selected for their capacity to securely contain gases for thousands to millions of years. These formations are typically located several kilometers below the Earth's surface and are covered by impermeable rock layers that act as natural seals, preventing any upward migration of stored CO_2. Decades of research, including pilot and commercial-scale projects, have shown that with proper site selection and engineering, the risks of leakage or environmental harm are extremely low. These findings are supported by organizations such as the International Energy Agency (IEA) and the Intergovernmental Panel on Climate Change (IPCC), which have endorsed CCS as a safe and essential climate mitigation tool.

Demonstrating Proven Track Records

Nothing builds confidence like real-world examples of success. Several CCS projects around the world have been operating safely for years, offering valuable case studies that can be used to illustrate the technology's effectiveness. The Sleipner project in Norway, which began in 1996, has stored over 20 million tons of CO_2 in a deep saline aquifer under the North Sea without any detectable leakage. Similarly, the Quest project in Canada and the Boundary Dam project in Saskatchewan have captured millions of tons of CO_2 from industrial facilities and safely stored it underground. These examples show that CCS is not a theoretical concept but a mature technology that can be implemented safely and effectively. Public outreach efforts should highlight these projects, emphasizing their environmental performance, safety records, and the benefits they have brought to surrounding communities.

Implementing Rigorous Monitoring and Verification Systems

One of the most effective ways to address concerns about CCS is to demonstrate the rigor and transparency of monitoring systems. Every phase of a CCS project, from site assessment to post-injection surveillance, involves detailed data collection and risk management. Technologies such as 3D seismic imaging, pressure sensors, gas tracers, and satellite monitoring allow for real-time tracking of CO_2 behavior in the subsurface. If any irregularities are detected, operators can take corrective action to prevent potential harm. Moreover, regulations in many jurisdictions require long-term monitoring even after a storage site has been sealed, ensuring continued oversight and environmental protection. Public access to monitoring data, regular reporting to regulatory agencies, and involvement of independent verifiers can further increase confidence in the safety of CCS operations.

Clarifying Legal and Regulatory Protections

Another important aspect of public concern revolves around liability and accountability. People want to know who is responsible if something goes wrong, and what mechanisms exist to protect health, property, and the environment. Robust regulatory frameworks are key to providing these assurances. Many countries have established legal guidelines that govern CCS projects, including requirements for permitting, environmental impact assessments, public consultations, and liability coverage. In the European Union, for instance, the CCS Directive outlines comprehensive rules for CO_2 storage safety, monitoring, and post-closure stewardship. In the United States, the Environmental Protection Agency (EPA) regulates underground injection under its Class VI well program, which includes extensive site characterization and risk analysis. Effective communication about these regulatory protections helps build trust and reassures the public that safety is not optional but mandatory.

Engaging Communities through Inclusive Dialogue

Top-down communication alone is not sufficient to address public concerns. Communities affected by CCS projects must be engaged as partners in the process, not just as passive recipients of information. Early, meaningful engagement allows community members to express their concerns, ask questions, and influence project design. This includes holding public meetings, forming community advisory panels, and providing platforms for ongoing dialogue. Project developers should make a genuine effort to understand local priorities, explain benefits and risks in a clear and respectful manner, and ensure that all voices are heard. When communities feel included and respected, they are more likely to support CCS initiatives or at least consider them with an open mind.

Addressing Ethical and Environmental Justice Issues

In some regions, public resistance to CCS stems from broader concerns about environmental justice. Communities that have historically borne the brunt of industrial pollution may view new infrastructure projects with suspicion, fearing further environmental degradation or health risks. These concerns are particularly acute in marginalized or low-income areas, where trust in authorities may be low. Addressing such concerns requires not only technical assurances but also ethical commitment. This includes ensuring that project benefits, such as jobs or infrastructure investment, are equitably distributed, that risks are minimized and transparently managed, and that affected populations have real power in decision-making. Incorporating principles of environmental justice into CCS planning and communication is essential for equitable and inclusive climate action.

Combating Misinformation and Misunderstandings

Misinformation can spread rapidly and undermine public trust, especially when it plays on fears of unknown or misunderstood technologies. Common misconceptions about CCS include the belief that stored CO_2 could explode, contaminate drinking water, or cause widespread health problems. While these scenarios are not supported by scientific evidence, they can take root if left unchallenged. Proactive communication campaigns should aim to counter misinformation with accurate, evidence-based information presented in accessible formats. Collaborating with trusted local leaders, educators, and scientists can help ensure that correct information reaches the public and is perceived as credible. Transparency, patience, and responsiveness are critical components of any effort to combat misinformation.

Positioning CCS within a Broader Climate Strategy

To be accepted, CCS must be seen as part of a holistic approach to climate change, not as a loophole for polluting industries. Communicators should emphasize that CCS complements renewable energy, energy efficiency, and behavioral change. It is particularly vital for decarbonizing sectors like cement and steel, where direct emissions are difficult to avoid. When paired with bioenergy (in BECCS) or used in direct air capture, CCS can also remove carbon from the atmosphere, helping to offset residual emissions and achieve net-negative targets. Framing CCS as one tool among many in a diverse climate toolkit can help mitigate the perception that it is a distraction or delay tactic.

5.5.3 Social License to Operate for CCS Projects

The success of Carbon Capture and Storage as a climate solution depends not only on technical excellence and regulatory compliance but also on social acceptance. The concept of social license to operate encapsulates the informal but essential approval of communities and the broader public. Achieving this license requires more than information; it demands genuine engagement, transparency, equity, cultural sensitivity, and long-term accountability. As the world moves urgently to implement climate solutions, building and maintaining social license for CCS projects must be a central strategy. Only by aligning technology with societal values can CCS fulfill its potential as a key pillar of the global response to climate change.

Social License

The term "social license to operate" originates from the mining and resource extraction sectors, where community opposition to environmentally disruptive projects became a frequent barrier to implementation. In the context of CCS, the social license encompasses public attitudes toward the safety, necessity, and fairness of capturing and storing carbon dioxide. It is influenced by factors such as environmental awareness, historical experiences with industrial projects, trust in institutions, and the inclusiveness of the decision-making process. A project with strong technical merits can still fail if it is perceived as imposed, secretive, or primarily serving corporate interests at the expense of local communities. Social license is not a one-time achievement; it requires continuous engagement, responsiveness to community concerns, and adaptive practices that align with evolving social expectations.

Importance of Community Engagement

Effective and early engagement with local communities is fundamental to establishing social license for CCS projects. Communities often hold valid concerns about potential risks, such as CO_2 leakage, induced seismicity, or long-term environmental consequences. These concerns can be exacerbated by misinformation or previous negative experiences with industrial activities. Therefore, project developers must prioritize dialogue over persuasion and listen actively to what residents value and fear. Meaningful engagement involves presenting technical information in accessible formats, offering forums for questions and feedback, and incorporating community input into project planning. This approach helps communities feel respected and empowered, fostering a sense of ownership and reducing resistance. When communities are treated as partners rather than obstacles, the pathway to social license becomes smoother and more sustainable.

Transparency and Trust as Cornerstones

Transparency is essential in earning and maintaining public trust, which is at the heart of social license. Project developers must openly share information about site selection criteria, risk assessments, environmental impact studies, monitoring plans, and mitigation strategies. Concealing data or failing to communicate uncertainties can quickly erode credibility. Trust is also built through consistency in communication and actions. When commitments are honored, and when feedback leads to observable changes, communities gain confidence in the integrity of the project and the people behind it. Independent third-party reviews and regular public updates on project performance and safety monitoring can further strengthen public trust. Trust takes time to build but can be lost quickly, which underscores the importance of sustained, honest communication throughout the project lifecycle.

Addressing Equity and Environmental Justice

Social license is not only about safety and information; it also involves fairness and equity. CCS projects should be designed and implemented in ways that consider the distribution of risks and benefits. Historically marginalized or economically disadvantaged communities may be more skeptical of CCS projects if they perceive them as extensions of extractive or polluting industries. Addressing environmental justice means ensuring that the burdens of CCS are not disproportionately borne by vulnerable populations and that these groups have equitable access to the economic and social benefits of the project. This can include job creation, local infrastructure improvements, and community development initiatives. Incorporating environmental justice principles into CCS planning can improve legitimacy and demonstrate that the project aligns with broader social and ethical goals.

Cultural Sensitivity and Local Context

The success of social license efforts often depends on cultural sensitivity and a deep understanding of the local context. What works in one community may not be effective in another. Factors such as local traditions, land use histories, economic dependencies, and political dynamics all shape how a CCS project is received. Indigenous communities, for example, may have unique legal rights and cultural ties to land that require specific forms of consultation and respect. Recognizing these distinct needs and adapting engagement strategies accordingly is crucial for gaining local approval. Social license is not a generic objective but a highly localized one, requiring tailored approaches that resonate with the specific values and concerns of the community involved.

Communication Strategies and Public Education

Public perception of CCS can be influenced by the way the technology is framed and communicated. Complex scientific information must be distilled into clear, relatable messages. Explaining how CCS contributes to climate mitigation, supports industrial jobs, and enhances environmental stewardship can help bridge the gap between abstract science and daily life. Visual aids, interactive platforms, site tours, and testimonials from other communities with successful CCS projects can enhance understanding and acceptance. Public education campaigns should not only target potential host communities but also reach broader national and regional audiences to build general awareness and support. A well-informed public is more likely to engage constructively and support climate solutions, including CCS.

Government and Policy

Government support is vital in legitimizing CCS projects and fostering social license. Clear regulatory frameworks, environmental safeguards, and transparent permitting processes provide a foundation of accountability and public assurance. In addition, government involvement can help mediate between developers and communities, ensuring that public interests are protected and grievances are addressed. Policy instruments such as financial incentives, community benefit agreements, and public engagement mandates can institutionalize best practices and encourage socially responsible development. When governments demonstrate that they prioritize safety, equity, and public participation, they help reinforce the conditions necessary for social license.

Responding to Opposition and Misconceptions

Even with proactive engagement and transparency, opposition to CCS projects may still arise. It is important to distinguish between uninformed skepticism and legitimate dissent. Addressing opposition requires patience, empathy, and a willingness to engage with critics. Project developers should avoid dismissing concerns and instead provide evidence-based responses that acknowledge risks and explain mitigation measures. Misconceptions about CCS, such as the idea that it is a tool to prolong fossil fuel use or that it poses grave environmental dangers, should be countered with scientific clarity and real-world examples. At the same time, developers must be prepared to adapt or even reconsider projects if community resistance proves insurmountable. Social license is ultimately granted by the community and cannot be forced.

Long-Term Commitment and Adaptive Management

Securing social license is not a one-off task achieved at the planning stage; it requires a long-term commitment to responsiveness and adaptation. As projects evolve and communities change, new issues may emerge that need to be addressed. Ongoing monitoring, transparent reporting, and open communication must continue throughout the life of a CCS project. Mechanisms such as community liaison offices, advisory boards, and feedback channels should remain active and accessible. Adaptive management allows developers to respond to new challenges and maintain the trust and support they have built. Demonstrating a continued commitment to the well-being of host communities helps ensure that the social license to operate remains intact over time.

CHAPTER

Environmental and Safety Considerations

6

6.1 ENVIRONMENTAL IMPACTS OF CCS TECHNOLOGIES

6.1.1 Risks of CO_2 Leakage and Groundwater Contamination

The risks of CO_2 leakage and groundwater contamination are among the most critical concerns associated with Carbon Capture and Storage. However, these risks are not insurmountable. Through meticulous site selection, advanced engineering practices, continuous monitoring, and strong regulatory frameworks, it is possible to minimize the likelihood of adverse outcomes. Moreover, transparent communication and public engagement are essential in building confidence in the technology. As climate change continues to pose existential threats, the importance of CCS will only grow. Addressing environmental risks proactively and responsibly will be key to ensuring that CCS can play a safe and effective role in the transition to a low-carbon future.

CO_2 Leakage in Geological Storage

Geological storage of CO_2 involves injecting the gas into porous rock formations deep beneath the Earth's surface, such as depleted oil and gas fields or saline aquifers. These formations are typically located more than 800 meters below ground, where pressures and temperatures ensure that CO_2 remains in a dense, supercritical state. For safe and effective storage, the injected CO_2 must remain trapped in these formations for thousands of years. However, the possibility of leakage exists if there are flaws in the geological caprock that seals the storage site or if abandoned wells or fractures act as conduits for gas migration. Leakage pathways can potentially lead CO_2 to migrate upwards through the rock layers and, in rare cases, reach the surface or contaminate overlying freshwater aquifers. The risk of such events depends on the site's geology, the integrity of the storage system, and the quality of operational practices.

Mechanisms of Groundwater Contamination

One of the key environmental risks associated with CO_2 leakage is the potential contamination of underground sources of drinking water. If CO_2 migrates from a storage formation into an overlying freshwater aquifer, it can dissolve in the water and lower its pH, making it more acidic. Acidified groundwater can react with minerals in the aquifer matrix, leading to the mobilization of trace metals and other contaminants such as arsenic, lead, or manganese. These substances, when released into the water supply, can pose serious risks to human health and ecological systems. In addition to chemical contamination, the physical displacement of groundwater due to CO_2 injection pressure could alter the natural flow paths of groundwater, affecting water availability and quality. However, the likelihood of such outcomes is relatively low when storage sites are properly selected and managed with stringent safety protocols.

Site Selection and Geological Suitability

One of the most effective ways to minimize the risks of CO_2 leakage and groundwater contamination is through careful site selection. Not all geological formations are suitable for CO_2 storage. An ideal site has a thick and impermeable caprock layer that can trap CO_2, a porous reservoir rock to store the gas, and a stable tectonic setting to minimize the risk of earthquakes or subsurface disturbance. In addition, the site should be located far below any potable aquifers and isolated from them by multiple geological barriers. Advanced geophysical and geochemical techniques are used to assess the integrity and suitability of potential storage sites, ensuring that only the safest locations are chosen for CO_2 injection. The extensive characterization process includes seismic surveys, core sampling, and modeling of fluid dynamics to predict how CO_2 will behave over time.

Engineering Controls and Injection Management

Engineering design and operational practices play a crucial role in preventing CO_2 leakage and protecting groundwater resources. Well construction must adhere to strict standards, with multiple layers of casing and cement to prevent the escape of CO_2 along the wellbore. During injection, operators must monitor pressure levels, injection rates, and subsurface conditions to ensure the system remains stable and within safe operational limits. Pressure management is particularly important to avoid fracturing the caprock or inducing seismic activity that could create leakage pathways. Additionally, inactive or legacy wells near the storage site are identified, evaluated, and remediated if necessary to prevent them from becoming potential conduits for CO_2 migration. These engineering controls, combined with real-time monitoring, create a robust safety net that reduces the risk of unintended environmental consequences.

Monitoring, Verification, and Risk Assessment

Continuous monitoring and verification are essential components of any CCS project. Sophisticated technologies are employed to track the movement of CO_2 underground and detect any anomalies that may indicate leakage. Techniques include time-lapse seismic imaging, pressure and temperature sensors, fluid sampling, and satellite-based surface monitoring. These data are used to validate predictive models and adjust operational parameters as needed. In addition to site-specific monitoring, risk assessments are conducted throughout the lifecycle of the project. These assessments evaluate potential hazards, the likelihood of various failure scenarios, and their potential consequences. The results inform the design of contingency plans and emergency response strategies, ensuring that operators are prepared to respond quickly and effectively should an issue arise.

Regulatory Frameworks and Legal Safeguards

Governments and regulatory bodies play a pivotal role in ensuring that CCS projects are implemented safely and with due consideration for environmental protection. In many jurisdictions, strict regulations govern every stage of a CCS project, from site selection and permitting to operation and post-closure monitoring. Regulatory agencies require developers to conduct environmental impact assessments, engage with stakeholders, and demonstrate that risks to groundwater and other resources have been thoroughly addressed. Post-injection responsibilities often include long-term stewardship, liability coverage, and ongoing monitoring to ensure that stored CO_2 remains securely trapped. These legal frameworks create an additional layer of assurance for the public and serve as a deterrent against negligence or non-compliance.

Public Perception and Communication Challenges

Despite the technical and regulatory safeguards in place, public concerns about CO_2 leakage and groundwater contamination remain a significant barrier to broader CCS deployment. Many people lack familiarity with the science behind geological storage and may perceive the technology as inherently risky or experimental. In some cases, opposition arises from fears that CCS will prolong the use of fossil fuels or lead to new forms of environmental degradation. To address these concerns, it is essential to engage with communities transparently and provide clear, evidence-based information about the risks and benefits of CCS. Demonstrating a track record of safe operations, involving independent experts, and creating opportunities for public participation can help build trust and alleviate fears.

Real-World Examples and Lessons Learned

Several operational CCS projects around the world provide valuable insights into managing the risks of CO_2 leakage and groundwater contamination. The Sleipner project in Norway, for example, has been storing CO_2 in a deep saline aquifer under the North Sea since 1996 without any reported leakage. The project employs extensive monitoring technologies and has become a model for international best practices. Similarly, the Quest project in Canada captures CO_2 from an oil sands facility and stores it in a secure underground formation, with comprehensive monitoring and public reporting protocols. These real-world experiences demonstrate that with rigorous site selection, careful engineering, and strong oversight, the environmental risks of CCS can be effectively managed.

6.1.2 Ecosystem Impacts of Ocean Storage

The ecosystem impacts of ocean storage of CO_2 are a critical consideration in the broader discussion of climate change mitigation. While this approach offers a potentially vast reservoir for carbon sequestration, it also poses serious risks to marine biodiversity, ecological processes, and the health of the oceans. Ocean acidification, disruption of food webs, hypoxia, and the vulnerability of deep-sea species are among the many concerns that must be addressed through rigorous scientific inquiry and careful regulation. As the world grapples with the urgent need to reduce greenhouse gas emissions, it is essential to ensure that the solutions pursued do not create new environmental problems. A comprehensive understanding of the ecological consequences of ocean storage is essential to making informed, responsible choices about its role in our climate future.

Ocean Storage and Its Mechanisms

Ocean storage of CO_2 primarily takes two forms: direct injection and natural dissolution. Direct injection involves transporting captured CO_2 and injecting it into deep ocean waters, usually at depths greater than 1,000 meters. At these depths, CO_2 behaves differently due to the high pressure and low temperature, forming either a dense plume that gradually dissolves or creating CO_2 hydrates that sink and remain suspended near the seabed. Alternatively, natural dissolution involves releasing CO_2 into mid-water columns where it can slowly mix and be absorbed by ocean currents over time. Both approaches aim to keep CO_2 isolated from the atmosphere for hundreds or thousands of years. However, they fundamentally alter the chemical composition of ocean water in localized areas, potentially affecting marine organisms and ecosystems.

Ocean Acidification and pH Changes

One of the most immediate impacts of ocean CO_2 storage is the lowering of pH in surrounding seawater. When CO_2 dissolves in water, it forms carbonic acid, which dissociates into bicarbonate and hydrogen ions. This process increases the acidity of the ocean and decreases the availability of carbonate ions, which are vital for marine organisms that rely on calcium carbonate to build shells and skeletons. Organisms such as corals, mollusks, and some plankton species are particularly vulnerable to pH changes. In areas near CO_2 injection sites, localized acidification can be severe, leading to physiological stress, reduced reproductive success, and even mortality in sensitive species. While deep-sea organisms are naturally adapted to higher pressure and lower temperatures, they are not necessarily tolerant to rapid changes in pH, making them susceptible to disruption from ocean storage activities.

Impacts on Deep-Sea Ecosystems

Deep-sea environments are some of the most poorly understood yet biologically rich ecosystems on Earth. These ecosystems include cold seeps, hydrothermal vents, abyssal plains, and trench communities that harbor unique life forms adapted to extreme conditions. Introducing CO_2 into these environments can alter the delicate balance that supports deep-sea biodiversity. Acidified waters may impact the metabolism, behavior, and survival of deep-sea fauna, especially species with limited mobility or narrow ecological niches. Additionally, the physical disturbance from CO_2 injection infrastructure and plumes may affect benthic habitats, leading to changes in species composition and potential declines in biodiversity. Because deep-sea species often have slow growth rates and low reproductive outputs, their recovery from disturbances can take decades or even centuries, raising ethical and ecological questions about the long-term impacts of ocean storage.

Disruption of Food Webs and Biological Productivity

Changes in pH and CO_2 concentrations in ocean waters can cascade through the food web, affecting both primary producers and higher trophic levels. For instance, phytoplankton, which form the base of the marine food chain, may experience altered growth patterns under changing chemical conditions. Some species may thrive in higher CO_2 environments, while others may decline, leading to shifts in species dominance and changes in nutrient cycling. Such imbalances can affect zooplankton populations, which in turn impact fish stocks and marine mammals. Ocean storage activities near productive fishing grounds or migratory routes could therefore disrupt commercial fisheries and the livelihoods of communities that depend on them. Moreover, some fish species use chemical cues to navigate, find food, and detect predators—abilities that could be impaired by acidified waters resulting from CO_2 storage.

Potential for Hypoxia and Altered Oxygen Dynamics

Another concern related to ocean storage is the potential for hypoxia, or low oxygen levels, in the water column. As CO_2-rich plumes dissolve and interact with surrounding water, they may increase microbial activity that consumes oxygen, especially near the seafloor. This can create or expand oxygen-depleted zones, which are already growing due to climate change and nutrient pollution. Hypoxic conditions are harmful to most marine life and can lead to "dead zones" where biodiversity is dramatically reduced. If CO_2 storage leads to the formation of such zones, the consequences for benthic ecosystems and nutrient cycling could be profound. Understanding and managing the risk of hypoxia is thus essential in evaluating the viability of ocean-based CCS strategies.

Uncertainty and Challenges in Predicting Long-Term Impacts

One of the major challenges in assessing the ecosystem impacts of ocean CO_2 storage is the inherent uncertainty in predicting how marine systems will respond over time. Ocean circulation patterns, biological interactions, and chemical processes are highly complex and vary across regions and seasons. While models can simulate the behavior of CO_2 plumes and pH changes, real-world conditions may lead to unexpected outcomes. Additionally, the cumulative effects of multiple stressors—such as warming temperatures, overfishing, and pollution—can interact with CO_2 storage impacts in unpredictable ways. Long-term ecological studies and pilot projects are needed to gather empirical data on how ocean ecosystems respond to CO_2 injection and to refine predictive models accordingly. Until such data is available, decision-makers must adopt a precautionary approach when considering ocean storage as a large-scale climate solution.

Legal and Ethical Considerations

Beyond the ecological impacts, the deployment of ocean storage raises important legal and ethical questions. International agreements such as the London Protocol and the United Nations Convention on the Law of the Sea (UNCLOS) regulate ocean dumping and marine pollution, and any ocean storage activity must comply with these frameworks. There is ongoing debate about whether deliberate CO_2 injection into the ocean constitutes pollution and whether it should be permitted under existing environmental laws. Ethically, the use of ocean ecosystems as carbon sinks introduces concerns about intergenerational equity and the rights of future generations to a healthy ocean. There are also concerns about the lack of informed consent from populations that may be affected by such projects, particularly coastal and indigenous communities whose livelihoods and cultural practices are tied to the sea. These considerations must be weighed alongside the technological feasibility of ocean storage to ensure that environmental justice and sustainability remain central to climate mitigation efforts.

Balancing Benefits and Risks

While ocean storage of CO_2 has the potential to contribute to global emission reduction targets, it must be carefully balanced against its ecological risks. The ocean is already under immense pressure from human activity, and introducing large volumes of CO_2 into marine environments may exacerbate existing problems. However, proponents argue that with proper site selection, monitoring, and adaptive management, ocean storage can be performed with minimal risk. This requires robust regulatory frameworks, transparent scientific assessment, and a commitment to ongoing research. Moreover, ocean storage should not be viewed as a substitute for reducing fossil fuel use or transitioning to renewable energy. Instead, it may play a role in addressing residual emissions that are otherwise difficult to eliminate.

6.1.3 Lifecycle Environmental Assessments of CCS Systems

Lifecycle environmental assessments provide a comprehensive and rigorous approach to evaluating the total environmental impact of Carbon Capture and Storage systems. By considering emissions, resource use, ecological effects, and infrastructure impacts across all phases of a project's life, LCA enables a deeper understanding of CCS's true environmental performance. While CCS holds promise for reducing industrial and energy sector emissions, it must be pursued in a way that maximizes benefits and minimizes unintended harm. Lifecycle analysis serves as a critical guide for achieving this balance, ensuring that CCS contributes meaningfully and responsibly to the global effort to stabilize the climate.

Lifecycle Perspective in CCS

Traditional assessments of CCS systems often focus on the carbon reduction achieved at the point of capture. However, this limited perspective can obscure the broader environmental consequences associated with the implementation of CCS. A lifecycle assessment evaluates all stages of a system's life, including material extraction, equipment manufacturing, facility construction, operation, maintenance, decommissioning, and post-closure monitoring. By adopting this holistic approach, LCA provides a more accurate understanding of the net environmental benefits or drawbacks of CCS technologies. This is especially important given the energy-intensive nature of some capture processes and the infrastructural demands of transport and storage. Through LCA, stakeholders can identify trade-offs, optimize processes, and make informed decisions about the deployment of CCS in various contexts.

Environmental Impacts of CO_2 Capture

The capture phase of CCS is often the most energy- and resource-intensive component of the system. Various technologies are used to separate CO_2 from flue gases, including post-combustion capture using chemical solvents, pre-combustion capture involving gasification, and oxy-fuel combustion. These methods typically require significant energy inputs, often sourced from the same fossil fuel plants that are being decarbonized. This added energy demand, known as the energy penalty, can lead to increased fuel consumption and upstream emissions, partially offsetting the benefits of captured CO_2. Moreover, the production and regeneration of chemical solvents can generate waste streams and secondary pollutants, such as nitrosamines or other degradation byproducts. The lifecycle impacts of these processes must be thoroughly examined to ensure that CO_2 capture does not introduce new environmental challenges.

Emissions and Resource Use in CO_2 Transport

Once captured, CO_2 must be transported to suitable storage sites, which can be located many kilometers away from the source. Transport is usually achieved through pipelines, though shipping and trucking may also be used in certain contexts. While pipeline transport is generally considered efficient and safe, its environmental impacts depend on several factors, including the distance to storage sites, terrain, and materials used in pipeline construction. LCAs assess emissions from steel manufacturing, land clearing, and energy use during pipeline operation. There are also potential risks associated with leaks during transport, which could pose safety and environmental hazards. Evaluating these impacts across the system's lifecycle helps to identify best practices for minimizing emissions, optimizing routes, and selecting the most environmentally sound transport methods.

Storage Phase and Long-Term Environmental Considerations

The final phase of CCS involves the injection of CO_2 into geological formations such as saline aquifers, depleted oil and gas reservoirs, or unmineable coal seams. While this phase is designed to offer long-term or permanent containment of CO_2, it must be assessed for both immediate and future environmental impacts. LCAs consider the energy requirements for injection, the construction of wells and monitoring equipment, and the potential effects on local ecosystems. Over the long term, storage sites must be monitored to ensure the integrity of the containment and to detect any signs of leakage. Although well-chosen and managed sites have demonstrated excellent retention capabilities, the risk of slow leakage over decades or centuries must be factored into lifecycle assessments. Additionally, the potential for induced seismicity or changes in groundwater flow should be analyzed, as these could have unforeseen consequences for local communities and environments.

Materials and Infrastructure Footprint

A significant portion of the environmental impact of CCS systems arises from the materials and infrastructure required for implementation. The construction of capture facilities, compressors, pipelines, and injection wells involves the use of concrete, steel, plastics, and other industrial materials. The extraction, processing, and transport of these materials contribute to greenhouse gas emissions, land use changes, and pollution. LCAs calculate the embodied energy and emissions associated with these inputs, offering a clearer picture of the full environmental cost of CCS deployment. Furthermore, maintenance activities, replacement of parts, and eventual decommissioning add to the cumulative footprint. Strategies to reduce the environmental burden of infrastructure—such as using recycled materials, extending component lifespans, and designing for disassembly—can be informed by lifecycle findings.

Comparative Analysis with Alternative Mitigation Strategies

Lifecycle assessments also allow for comparison between CCS and other climate mitigation strategies, such as renewable energy deployment, energy efficiency improvements, and nature-based solutions like afforestation. While CCS can play a vital role in decarbonizing hard-to-abate sectors like cement and steel production, it may not always offer the most cost-effective or environmentally beneficial solution in every case. LCAs can reveal situations where the lifecycle emissions of CCS, particularly in poorly managed or inefficient systems, approach or even exceed those of the original unabated process. Conversely, when implemented efficiently, CCS can provide substantial net reductions in CO_2 emissions, especially in applications where no other low-carbon alternatives are currently viable. By comparing lifecycle performance, policymakers can allocate resources and incentives more effectively across different mitigation pathways.

Incorporating Biodiversity and Ecosystem Impacts

While LCAs traditionally focus on quantifiable environmental metrics like greenhouse gas emissions, energy use, and pollution, there is growing recognition of the need to consider broader ecological impacts, including effects on biodiversity and ecosystems. CCS projects may alter land use patterns, disturb habitats, or affect water availability and quality. For example, large-scale deployment of CCS may compete for space with agriculture, forestry, or conservation areas. Infrastructure development, such as pipeline networks, can fragment habitats or introduce invasive species. LCAs that incorporate biodiversity and land-use change indicators provide a more nuanced understanding of CCS impacts and help ensure that emission reductions do not come at the expense of ecological integrity.

Challenges and Uncertainties in Lifecycle Assessments

Conducting LCAs of CCS systems presents several methodological challenges. Data availability can be limited, especially for emerging technologies or pilot projects. Assumptions about system boundaries, timeframes, and technological performance can significantly influence results. Additionally, uncertainty in future developments—such as advances in capture efficiency, energy mix evolution, or storage site behavior—complicates long-term projections. Despite these limitations, LCA remains a powerful tool when conducted transparently and with sensitivity analyses. The growing development of standardized LCA methodologies and databases specific to CCS helps improve consistency and comparability across studies.

Policy Implications and Role in Decision-Making

The insights provided by lifecycle assessments are essential for shaping effective climate policies and regulatory frameworks. Governments and funding bodies can use LCA results to prioritize projects that deliver the highest environmental return on investment. Lifecycle data can also inform carbon pricing schemes, subsidies, and reporting requirements, ensuring that CCS projects are held to rigorous environmental standards. Moreover, LCA findings can support transparency and public acceptance by demonstrating the net benefits of CCS and clarifying potential trade-offs. As climate goals become more ambitious, integrating LCA into policy design helps ensure that mitigation measures deliver not only carbon reductions but also broader environmental sustainability.

6.2 SAFETY STANDARDS FOR CCS OPERATIONS

6.2.1 Risk Assessment and Mitigation Strategies

Risk assessment and mitigation strategies are foundational to the successful deployment of Carbon Capture and Storage technologies. Given the complexity and potential hazards associated with CCS, it is critical to conduct thorough and site-specific risk assessments that address technical, environmental, societal, and financial dimensions. By identifying potential risks early and developing robust mitigation plans, stakeholders can enhance the safety, reliability, and public acceptance of CCS projects. These strategies ensure that CCS can contribute effectively to global decarbonization efforts while minimizing unintended consequences and building long-term sustainability and trust in the technology.

Comprehensive Risk Assessment

Risk assessment in the context of CCS involves identifying potential hazards throughout the entire lifecycle of a project, estimating the likelihood of their occurrence, and evaluating their potential consequences. These risks may arise during CO_2 capture, compression, transportation, injection, or long-term storage phases. A comprehensive risk assessment must be based on scientific data, site-specific conditions, and probabilistic modeling. This process helps to determine the safety of proposed projects, comply with regulatory requirements, and build public confidence. Risk assessments also provide a framework for continuous monitoring and improvement by highlighting areas where additional data, mitigation, or technology upgrades are required.

Risks Associated with CO_2 Capture

During the capture phase, one of the main risks is related to the operational reliability and energy intensity of the technology used. Capture technologies such as amine-based solvents or membrane systems can introduce chemical hazards, including the release of toxic degradation byproducts. In addition, increased energy consumption for capture can lead to higher fuel use, which in turn may increase emissions of other pollutants such as nitrogen oxides or particulate matter unless mitigated. Operational failures such as leaks, spills, or equipment malfunctions can pose risks to workers and the surrounding environment. These risks need to be assessed through hazard identification, material safety evaluations, and operational reliability studies.

Transport Risks and Mitigation

Transporting CO_2, primarily via pipelines, introduces another layer of risk, particularly due to the high-pressure conditions under which CO_2 is conveyed. A rupture or leak in a pipeline could lead to the sudden release of large volumes of CO_2, which can displace oxygen in the atmosphere and pose suffocation risks to humans and animals. Moreover, CO_2 is an asphyxiant and, in high concentrations, can cause unconsciousness or death without proper warning. The risk assessment process must evaluate factors such as pipeline integrity, route selection, population exposure, and emergency response readiness. Mitigation strategies include using corrosion-resistant materials, regular inspections, smart sensors for leak detection, and establishing exclusion zones and safety protocols around pipelines.

Risks during Injection and Storage

Injecting CO_2 into underground geological formations poses several geological and engineering risks. These include well integrity failures, unanticipated migration of CO_2, induced seismicity, and caprock breach. Well failure can lead to leaks that may release CO_2 to the surface or into underground sources of drinking water. Poorly understood subsurface geology can result in CO_2 escaping the intended storage formation and moving into adjacent rock layers.

Induced seismicity, or minor earthquakes triggered by the pressure changes associated with CO_2 injection, can compromise the structural integrity of the reservoir and surrounding formations. Effective risk assessment must therefore include detailed geological surveys, modeling of reservoir behavior, and historical seismic data analysis. Mitigation involves site selection based on robust geological stability, the use of multiple sealing layers, proper well construction practices, and pressure management strategies.

Environmental and Ecosystem Risks

The environmental risks of CCS also extend to potential impacts on ecosystems, especially in the event of accidental CO_2 releases or contamination of groundwater. CO_2 leakage into shallow aquifers can lead to the mobilization of toxic metals or changes in pH that affect water quality. Additionally, surface leaks could harm plant life or wildlife in the vicinity. Lifecycle environmental impacts, including those from construction and decommissioning activities, also need to be evaluated. Mitigation strategies in this context involve continuous environmental monitoring, protective measures for sensitive habitats, and rapid-response plans for environmental remediation in case of leakage or contamination.

Monitoring and Verification Systems

One of the most effective strategies for mitigating risks in CCS is the implementation of comprehensive monitoring and verification systems. These systems track the behavior of CO_2 in the subsurface, detect anomalies, and provide early warnings in case of potential issues. Monitoring technologies include seismic surveys, pressure sensors, geochemical sampling, and satellite imaging. Real-time monitoring data can inform adaptive management practices and help operators respond swiftly to deviations from expected performance. Verification processes also ensure that stored CO_2 remains securely contained and that projects comply with regulatory and environmental standards. Transparent reporting of monitoring data enhances public trust and ensures accountability.

Societal and Regulatory Risks

Beyond technical and environmental risks, CCS projects face social and regulatory risks that can impede project approval and implementation. Public resistance, often rooted in concerns about safety, environmental impacts, or lack of trust in operators, can lead to delays or cancellations. In some cases, local communities may not accept projects due to insufficient engagement or perceived imposition of risks without adequate benefits. Regulatory uncertainty, especially in jurisdictions without established CCS frameworks, can create financial and operational risks for investors and project developers. To mitigate these issues, it is essential to engage with stakeholders early and often, ensure transparency in risk communication, and develop clear regulatory pathways that define responsibilities, liabilities, and standards.

Risk Mitigation through Site Selection and Design

Selecting appropriate sites for CO_2 storage is one of the most critical risk mitigation strategies. Ideal sites should possess stable geological formations, impermeable caprocks, and minimal risk of seismic activity. In addition, proximity to CO_2 sources can reduce transportation risks and costs. Site characterization involves detailed geological mapping, sampling, and simulation of subsurface behavior to predict the movement and fate of injected CO_2. Proper design of injection wells and infrastructure also minimizes the likelihood of mechanical failures or leakage.

Incorporating safety margins in pressure management, injection rates, and reservoir capacity further enhances system resilience.

Emergency Preparedness and Response Plans

Despite best efforts, unexpected events may still occur during CCS operations. Therefore, emergency preparedness and response planning are essential components of a comprehensive risk mitigation strategy. These plans must outline clear procedures for detecting, responding to, and containing CO_2 releases or other incidents. Training of personnel, coordination with local emergency services, and public awareness campaigns form the backbone of effective emergency response. Periodic drills, maintenance of safety equipment, and communication protocols ensure that responses are swift and effective, minimizing harm to people and the environment.

Economic and Financial Risk Management

Financial risks also play a role in CCS implementation, including cost overruns, uncertain carbon markets, and long-term liabilities. From a risk assessment perspective, these concerns must be addressed through robust project planning, cost-benefit analysis, and the development of insurance or financial assurance mechanisms. Governments and industry stakeholders can reduce financial uncertainty by establishing long-term carbon pricing signals, subsidies, and legal frameworks for liability management. Mitigation of financial risks also requires a clear understanding of operational risks and the adoption of best practices to avoid costly disruptions or failures.

6.2.2 Emergency Response and Contingency Plans

Emergency response and contingency planning are indispensable components of safe and sustainable Carbon Capture and Storage deployment. These plans provide a structured framework for anticipating, managing, and recovering from emergencies that may arise during the capture, transport, and storage of CO_2. By incorporating hazard identification, response procedures, public engagement, and coordination with authorities, CCS operators can protect both people and the environment. Robust emergency planning also plays a key role in regulatory compliance and community acceptance, thereby supporting the long-term success and credibility of CCS as a climate mitigation solution. As the technology evolves and scales, continuous improvement in emergency preparedness will remain a cornerstone of responsible CCS development.

Need for Emergency Response Planning

Emergency response planning for CCS projects is rooted in the acknowledgment that despite rigorous risk assessments and mitigation strategies, unforeseen incidents may occur. These incidents may include accidental CO_2 releases from pipelines, well blowouts during injection, equipment failures, or geophysical disturbances such as induced seismicity. Unlike conventional natural gas leaks, CO_2 leaks present unique hazards due to the gas's asphyxiating properties at high concentrations. Therefore, emergency response plans must be specifically designed to manage the unique characteristics of CO_2 behavior in both confined and open environments. These plans are critical for safeguarding workers, nearby communities, and ecosystems and for ensuring the long-term viability and public acceptance of CCS projects.

Components of a Robust Emergency Response Plan

A comprehensive emergency response plan in the context of CCS encompasses several core components that work together to create a rapid and effective reaction to emergencies. These include hazard identification, risk analysis, response procedures, communication protocols, training programs, and coordination with local authorities. Hazard identification and risk analysis help to anticipate potential emergency scenarios and evaluate their severity and likelihood. Response procedures outline the steps to be taken in various emergency situations, such as isolating leaks, evacuating personnel, deploying containment equipment, and conducting environmental remediation. Communication protocols ensure timely notification of all stakeholders, including workers, regulators, emergency services, and the public. Regular training and drills enhance the readiness of personnel, and coordination with local emergency responders ensures that external support can be mobilized quickly when necessary.

Emergency Preparedness for CO_2 Capture Facilities

At the capture stage, emergency response planning focuses primarily on managing operational hazards such as fires, chemical spills, and mechanical failures. CO_2 capture processes often involve the use of solvents, high-pressure equipment, and complex thermal systems. An emergency at this stage might involve the release of toxic chemicals, pressure-related explosions, or fire outbreaks. Emergency response plans for capture facilities must include fire suppression systems, spill containment kits, personal protective equipment (PPE), and procedures for immediate plant shutdown. Emergency exits, alarm systems, and first-aid facilities should be clearly marked and readily accessible. Worker training programs should emphasize early detection of malfunctions, proper use of safety gear, and prompt evacuation protocols.

Responding to Transportation Incidents

Transportation of CO_2, mainly via pipelines but sometimes by road or sea, introduces specific emergency challenges. CO_2 is transported in a supercritical state, under high pressure and low temperature. A sudden release due to pipeline rupture or valve failure can lead to a dense gas cloud that may travel along the ground, potentially causing asphyxiation. Emergency response plans must account for the physical behavior of CO_2 and consider meteorological conditions, topography, and population proximity. Pipeline monitoring systems should be complemented by shut-off valves, emergency pressure relief systems, and automated leak detection technologies. In the event of a leak, responders should be trained to identify danger zones, evacuate affected areas, and use gas dispersion models to predict the movement of CO_2 clouds. Coordination with local emergency services, including fire and medical responders, is vital to ensure that response efforts are timely and effective.

Handling Emergencies during CO_2 Injection and Storage

The injection and storage phase of CCS carries risks such as well blowouts, casing failures, and unintended migration of CO_2 to the surface or into aquifers. These scenarios can lead to the uncontrolled release of CO_2 or contamination of groundwater sources. Emergency response plans for storage sites must address well integrity monitoring, pressure management, and rapid sealing techniques. Blowout preventers and shut-in valves are essential for halting uncontrolled releases during drilling or injection. Storage facilities should be equipped with downhole pressure gauges, microseismic sensors, and surface gas detectors. In the event of a breach, responders must be able to quickly isolate the injection well, contain the release, and assess the extent of subsurface migration using geophysical and geochemical tools. Groundwater monitoring programs should be integrated into contingency planning to detect and respond to contamination early.

Public Safety and Community Engagement

Ensuring public safety is a critical dimension of any emergency response strategy. CCS projects are often situated near communities that may be unaware of the risks associated with CO_2 infrastructure. Transparent communication about the safety features of CCS and what to do in the event of an emergency helps build trust and preparedness. Public awareness campaigns should include clear information on evacuation routes, warning signals, and emergency contact numbers. CCS operators should maintain a public emergency notification system, such as sirens, SMS alerts, or radio announcements, to quickly inform residents of incidents. Collaboration with local schools, community centers, and healthcare facilities strengthens the preparedness network and ensures that vulnerable populations are considered in contingency plans.

Training and Simulation Drills

Effective response to emergencies relies on the readiness and competence of personnel involved. Regular training programs for CCS staff are crucial for maintaining high safety standards. These programs should cover hazard recognition, emergency shutdown procedures, first aid, fire control, and use of safety equipment. Simulation drills allow workers and emergency responders to practice coordinated responses to different emergency scenarios. These drills test the functionality of equipment, the clarity of communication protocols, and the speed of evacuation or containment measures. Lessons learned from drills can be used to revise and improve response plans. Participation by local emergency services in these exercises further enhances inter-agency coordination and ensures that all stakeholders are aligned.

Coordination with Regulatory and Emergency Agencies

CCS projects must operate within a framework of national and regional safety regulations. Emergency response plans should be developed in consultation with regulatory authorities and aligned with relevant laws and standards. Collaboration with emergency management agencies, environmental protection bodies, and health departments ensures that response strategies are comprehensive and legally compliant. Regular audits and inspections by regulators help validate the adequacy of emergency plans and identify areas for improvement. Establishing joint emergency response committees or working groups fosters an integrated approach to preparedness, allowing for shared resources, expertise, and responsibilities.

Post-Incident Recovery and Remediation

An effective emergency response plan does not end with the containment of an incident; it must also include provisions for post-incident recovery and remediation. This includes environmental cleanup, assessment of long-term impacts, repair of damaged infrastructure, and support for affected communities. Environmental monitoring should continue after the incident to evaluate recovery progress and detect any residual contamination. Transparent reporting of the incident, investigation findings, and corrective actions helps restore public trust and contributes to broader knowledge in CCS safety management. Lessons learned from incidents must be incorporated into future planning and used to update training, design standards, and operational practices.

6.2.3 Occupational Health and Safety in CCS Facilities

Occupational health and safety in CCS facilities is a multidimensional challenge that requires vigilance, planning, and commitment at every level of operation. From chemical hazards and mechanical risks to emergency response and health monitoring, CCS operators must implement comprehensive OHS systems that are adapted to the specific characteristics of carbon capture, transport, and storage. A strong safety culture, supported by training, technology, and regulatory oversight, ensures that the workforce remains protected while enabling CCS projects to contribute meaningfully to climate mitigation. As CCS technology matures and expands, prioritizing worker safety will be essential for achieving both operational excellence and sustainable environmental progress.

Occupational Risks in CCS

CCS systems encompass a series of interconnected processes that include capturing CO_2 from emission sources, compressing and transporting it, and injecting it into deep geological formations for long-term storage. Each of these stages presents unique occupational risks. During the capture process, workers may be exposed to hazardous chemicals such as amines, solvents, or other reagents used to extract CO_2 from flue gases. The compression and transportation phases involve high-pressure equipment and pipelines, which pose risks of mechanical failure, high-pressure gas release, and physical injury. At storage sites, potential hazards include exposure to high-pressure CO_2, wellhead accidents, and complications from drilling or maintenance operations. A comprehensive approach to occupational safety must therefore address risks across the entire CCS lifecycle.

Chemical Hazards and Exposure Prevention

One of the most pressing health concerns in CCS facilities is chemical exposure, particularly during the capture phase where solvents and other reagents are handled. These chemicals can be toxic or irritating, and in some cases, they may degrade into hazardous byproducts under high temperature and pressure. Long-term exposure may result in respiratory issues, skin conditions, or other chronic health problems. To prevent such risks, CCS facilities must enforce strict handling procedures, provide appropriate personal protective equipment (PPE), and implement closed-loop systems that minimize the release of vapors. Monitoring air quality in work areas is critical, as is training workers to recognize symptoms of exposure and to use chemical materials safely. Safety data sheets (SDS), proper labeling, and emergency wash stations are essential elements of workplace preparedness.

Mechanical and Process Safety

The operation of CCS infrastructure involves high-pressure systems, rotating machinery, and thermal processes, all of which present mechanical hazards. Accidents may include burns, lacerations, crush injuries, or equipment-related fatalities if safety measures are not properly enforced. Process safety in CCS facilities requires rigorous design standards, regular maintenance schedules, and automation systems that include fail-safes and shutdown mechanisms.

Workers must be trained in lockout-tagout (LOTO) procedures, which ensure that machines are properly shut down and cannot be restarted until maintenance is complete. Adequate spacing, shielding, and ventilation in mechanical rooms are also necessary to prevent injuries. Mechanical integrity programs should include routine inspections, non-destructive testing, and vibration analysis to identify and mitigate wear-and-tear risks before they escalate into incidents.

Handling Pressurized CO_2 and Leak Risks

Carbon dioxide, while not flammable or explosive, is hazardous in high concentrations. As a colorless and odorless gas, CO_2 can displace oxygen and cause asphyxiation without warning. This risk is especially high in confined spaces such as tunnels, control rooms, or depressions where the gas may accumulate. High-pressure pipelines and injection wells used in CCS projects further elevate the potential for dangerous CO_2 releases. To protect workers, safety systems must include continuous CO_2 monitoring, ventilation systems, and alarm mechanisms that alert personnel to dangerous levels. Workers must also be trained in recognizing early signs of CO_2 exposure such as dizziness, shortness of breath, and confusion. Emergency protocols should be in place to evacuate personnel swiftly and administer medical aid if needed.

Drilling, Injection, and Well Integrity Risks

Storage of CO_2 in geological formations requires the use of wells for injection, monitoring, and sometimes maintenance activities. Drilling and well operations introduce occupational hazards including blowouts, exposure to drilling fluids, high-pressure injuries, and falls from height. These risks are similar to those found in the oil and gas industry, and best practices from that sector can be applied to CCS. Ensuring well integrity through proper casing and cementing is critical not only for environmental protection but also for worker safety. Blowout preventers (BOPs), proper well control training, and the use of safety harnesses and anti-slip platforms are vital measures to minimize occupational hazards during drilling and injection operations.

Workplace Ergonomics and Fatigue Management

Beyond acute physical risks, CCS facilities must also consider long-term occupational health challenges such as musculoskeletal disorders and fatigue. Workers who operate heavy equipment, perform repetitive tasks, or spend long hours on site may develop chronic conditions if ergonomics are not properly addressed. Ergonomically designed control rooms, adjustable workstations, and mechanical aids for lifting and carrying can significantly reduce these risks. Additionally, rotating shift schedules, adequate break times, and stress management programs help combat fatigue and promote mental well-being. Fatigue in high-risk environments like

CCS operations can lead to lapses in judgment or slower reaction times, making effective fatigue management a crucial aspect of occupational safety.

Emergency Preparedness and Safety Culture

A robust safety culture is at the heart of any effective OHS system. In CCS facilities, emergency preparedness must be a routine part of workplace operations. This includes conducting regular emergency drills, maintaining first-aid stations, and ensuring that all employees are familiar with evacuation routes and communication systems. Safety training should be a recurring activity that keeps workers updated on new procedures, equipment, and safety protocols. The development of a safety-first mindset, supported by leadership and reinforced through incentives and open communication, can significantly reduce incident rates. Incident reporting systems should be non-punitive and designed to capture near-misses as well as accidents, enabling a proactive approach to safety management.

Health Surveillance and Medical Monitoring

To protect workers over the long term, CCS facilities must implement health surveillance programs that monitor for signs of chronic illness or exposure-related conditions. Regular medical check-ups, respiratory function tests, and screenings for chemical sensitivity or hearing loss are valuable tools in early detection and prevention. These programs should be confidential and include counseling services to support both physical and mental health. Health data should inform workplace adjustments, such as changes in shift patterns, job assignments, or protective equipment. In high-risk areas, wearable health monitors can also be used to track vital signs and exposure levels in real time, enhancing the ability to respond to emergencies swiftly.

Regulatory Compliance and Industry Standards

OHS regulations for CCS projects are governed by national workplace safety laws and industry-specific guidelines. These include standards from organizations such as the Occupational Safety and Health Administration (OSHA), the International Labour Organization (ILO), and relevant environmental and energy authorities. Compliance involves documentation, auditing, and performance reporting. CCS operators must maintain detailed records of safety inspections, incident investigations, worker certifications, and health monitoring data. Regular audits and inspections help identify weaknesses and ensure that all procedures meet or exceed regulatory requirements. Non-compliance not only endangers workers but also jeopardizes project viability and public trust.

6.3 CARBON UTILIZATION AND REUSE

6.3.1 Turning CO_2 into Valuable Products: Carbon Conversion Technologies

Turning CO_2 into valuable products through carbon conversion technologies offers a visionary yet increasingly tangible solution to the climate crisis. These technologies provide an opportunity to close the carbon loop, mitigate emissions, and generate economic value from waste gases. By harnessing advances in chemistry, biology, and engineering, and by aligning policy and investment strategies, societies can transition toward a more sustainable and resilient industrial future. While significant challenges remain, the growing momentum around carbon utilization underscores its potential as a key pillar of the low-carbon economy.

Scientific Basis for Carbon Conversion

Carbon conversion technologies rely on the chemical transformation of CO_2, a thermodynamically stable molecule, into more reactive and useful forms. Because CO_2 is an inert gas under standard conditions, significant energy input or advanced catalysis is required to break its chemical bonds and rearrange them into new compounds. Various processes have been developed to facilitate this transformation, including thermochemical, electrochemical, photochemical, and biological methods. The success of these methods depends on the availability of low-carbon energy, cost-effective catalysts, and scalable infrastructure. Innovations in materials science, nanotechnology, and enzyme engineering are driving rapid advancements in this field, making carbon conversion a promising area of climate technology.

Electrochemical Reduction of CO_2

Electrochemical reduction is one of the most promising avenues for CO_2 conversion. In this method, electricity—ideally sourced from renewable energy—is used to drive chemical reactions that convert CO_2 into carbon monoxide, formic acid, methane, ethylene, or other hydrocarbons. These products can be used as industrial feedstocks, energy carriers, or fuel additives. The process typically involves an electrolyzer with a cathode where CO_2 is reduced and an anode where oxidation occurs. The efficiency and selectivity of the reaction depend heavily on the catalyst material. Recent research has explored copper-based catalysts for their ability to produce multi-carbon products, while silver and tin have shown promise in generating formate. Although challenges remain in scaling up and reducing costs, electrochemical CO_2 reduction holds great potential due to its compatibility with decentralized and renewable energy sources.

Thermochemical and Catalytic Pathways

Thermochemical conversion involves the use of high temperatures and catalysts to transform CO_2 into valuable chemicals. One of the best-known processes is dry reforming, where CO_2 reacts with methane to produce synthesis gas (syngas), a mixture of hydrogen and carbon monoxide that serves as a building block for fuels and chemicals. Another significant process is methanation, in which CO_2 is converted into methane using hydrogen. This reaction, known as the Sabatier process, has garnered attention for its potential to produce synthetic natural gas. Catalysts such as nickel, ruthenium, or cobalt are essential for enhancing reaction rates and lowering energy requirements. Although thermochemical methods often require substantial energy input, integrating them with waste heat recovery or renewable thermal energy can improve overall efficiency and sustainability.

Photocatalytic and Solar-Driven Conversion

Photocatalytic conversion of CO_2 seeks to mimic the natural process of photosynthesis by using sunlight to drive chemical reactions. In this approach, semiconductors or photoactive materials are exposed to light, creating electron-hole pairs that facilitate the reduction of CO_2 and the oxidation of water. Products of these reactions can include methane, methanol, or other hydrocarbons. This technology has the potential to create solar fuels—clean energy carriers derived from CO_2 and sunlight. However, the field is still largely in the research and development phase, as efficiencies remain low and long-term catalyst stability is an issue. Nonetheless, with improvements in light-harvesting materials and reactor design, photocatalytic CO_2 conversion could emerge as a viable and carbon-neutral energy solution.

Biological and Biotechnological Approaches

Nature has long evolved mechanisms to fix CO_2 through photosynthesis and microbial metabolism. Leveraging these natural processes, researchers are developing biotechnological solutions to convert CO_2 into valuable products using algae, bacteria, or genetically engineered microbes. Algae cultivation systems, for instance, can use captured CO_2 to grow biomass, which is then processed into biofuels, animal feed, or bioplastics. Similarly, certain strains of microbes can metabolize CO_2 and hydrogen to produce ethanol, acetate, or other organic acids. Advances in synthetic biology have enabled the creation of custom microbial strains designed to produce specific high-value chemicals from CO_2. These systems offer the advantage of operating under mild conditions and can be integrated with industrial CO_2 sources such as fermentation plants or power stations.

Concrete and Building Materials from CO_2

One of the most commercially advanced applications of CO_2 conversion lies in the production of building materials. CO_2 can be mineralized to form stable carbonates, which are then used to produce concrete, bricks, or aggregates. In some processes, CO_2 is injected into wet concrete mixtures, where it reacts with calcium ions to form calcium carbonate, enhancing the strength of the final product. This process not only sequesters CO_2 permanently but also reduces the carbon footprint of one of the most emissions-intensive industries: cement and construction. Companies across the globe are scaling up carbon-based concrete technologies, supported by policies and green building initiatives that promote low-carbon infrastructure.

Synthetic Fuels and Industrial Applications

Converting CO_2 into synthetic fuels, or e-fuels, represents an exciting frontier in the effort to decarbonize sectors such as aviation, shipping, and heavy industry, which are difficult to electrify. These fuels are produced by combining CO_2 with green hydrogen through catalytic processes to create hydrocarbons such as methanol, dimethyl ether, or synthetic diesel. These fuels are chemically similar to their fossil-derived counterparts and can be used in existing engines and infrastructure. The advantage of synthetic fuels is their potential to offer a closed carbon loop, where emissions from combustion are captured and reused. However, producing these fuels is energy-intensive and currently more expensive than fossil fuels, making further innovation and policy support essential for competitiveness.

Economic and Environmental Considerations

While the technical feasibility of CO_2 conversion is improving, economic viability remains a challenge. Most CO_2-derived products are currently more expensive than their conventional alternatives due to high energy demands, low conversion efficiencies, and limited economies of scale. To become commercially competitive, these technologies require continued research, favorable regulatory frameworks, carbon pricing mechanisms, and investment in demonstration projects. On the environmental side, the carbon intensity of the entire conversion process must be considered. If powered by fossil energy, the climate benefits may be nullified. Therefore, coupling CO_2 conversion technologies with renewable energy sources is vital to ensuring that the process remains carbon-negative or at least carbon-neutral.

Policy and Market Incentives for Carbon Utilization

Government policies play a critical role in accelerating the adoption of CO_2 conversion technologies. Carbon pricing, tax credits for CO_2 utilization, renewable energy mandates, and funding for pilot projects can create favorable market conditions. Programs like the U.S. 45Q tax credit, the European Union's Innovation Fund, and national low-carbon fuel standards provide financial incentives that help bridge the gap between research and commercialization. In parallel, corporate sustainability goals and consumer demand for carbon-neutral products are encouraging industries to explore carbon utilization pathways as part of their decarbonization strategies.

6.3.2 Synergies with Renewable Energy: Hydrogen and Synthetic Fuels

The synergy between renewable energy, hydrogen, and synthetic fuels represents a powerful pathway toward a decarbonized energy future. These technologies complement intermittent renewables by providing storage solutions, enabling carbon recycling, and extending clean energy benefits to hard-to-electrify sectors. While significant technical, economic, and infrastructural challenges remain, the momentum around hydrogen and e-fuels is growing rapidly. With continued innovation, investment, and policy support, these technologies can play a central role in achieving global climate goals and transforming the energy landscape into one that is sustainable, secure, and inclusive.

Green Hydrogen in Renewable Energy Integration

Hydrogen, particularly green hydrogen produced through the electrolysis of water using renewable electricity, acts as a bridge between intermittent renewable sources and consistent energy supply. Unlike fossil fuels, green hydrogen emits no carbon dioxide when used, and its production utilizes surplus renewable energy that might otherwise go to waste. As wind and solar power become more prominent in the energy mix, the challenge of managing variable output becomes significant. Hydrogen offers a solution by serving as a form of energy storage. Excess electricity generated during peak sunlight or strong wind conditions can power electrolysis systems that split water into hydrogen and oxygen. The hydrogen can then be stored for long durations and used when renewable generation is low, effectively stabilizing the grid and enhancing energy resilience.

Hydrogen as a Feedstock for Synthetic Fuels

Beyond its direct use as a clean fuel, hydrogen serves as a key feedstock for the production of synthetic fuels. When combined with captured CO_2, green hydrogen can be converted into hydrocarbons that are chemically identical to conventional fossil fuels. These synthetic fuels, often referred to as e-fuels, can be used in existing internal combustion engines, aircraft turbines, and industrial furnaces without requiring major infrastructure changes. The primary processes involved include methanol synthesis, Fischer-Tropsch synthesis, and methanation. Each of these relies on hydrogen and CO_2 as input materials and results in liquid or gaseous fuels such as synthetic diesel, kerosene, or methane. The integration of renewable hydrogen with carbon utilization pathways makes synthetic fuels a viable solution for achieving net-zero emissions in sectors where electrification is not yet feasible.

Decarbonizing Hard-to-Electrify Sectors

One of the most compelling arguments for developing hydrogen and synthetic fuel technologies lies in their ability to decarbonize hard-to-abate sectors. Heavy-duty transportation, maritime shipping, and aviation all require high energy density fuels that batteries currently cannot provide efficiently. Similarly, certain industrial processes, such as steel and cement production, involve high-temperature heat and chemical reactions that are difficult to achieve using electricity alone. Hydrogen can replace coke in steelmaking through direct reduction, and it can fuel kilns in cement factories. Synthetic fuels offer a drop-in alternative for jet fuel and marine diesel, making them particularly attractive for aviation and shipping, which currently have limited decarbonization options. These applications not only reduce greenhouse gas emissions but also decrease reliance on imported fossil fuels, enhancing national energy security.

Hydrogen Storage and Distribution Challenges

Despite its potential, hydrogen presents significant challenges in terms of storage, transport, and distribution. Hydrogen has a low volumetric energy density, requiring it to be stored at high pressures, low temperatures, or in chemical compounds such as ammonia or metal hydrides. Each storage method involves trade-offs between cost, safety, and efficiency. Additionally, hydrogen is highly flammable and can cause material embrittlement in pipelines, necessitating careful selection of materials and robust safety protocols. Developing a hydrogen infrastructure that includes production facilities, storage units, pipelines, and refueling stations requires substantial investment and coordinated planning. However, advancements in compression technologies, liquefaction, and chemical carriers are gradually addressing these barriers, making hydrogen more accessible as a large-scale energy carrier.

Synthetic Fuels as Energy Carriers and Carbon Recyclers

Synthetic fuels provide a unique advantage by serving as both energy carriers and mechanisms for carbon recycling. Unlike hydrogen, synthetic fuels are compatible with existing logistics and fueling infrastructure, including pipelines, tankers, and fueling stations. This compatibility makes them an attractive option for countries and industries looking to reduce emissions without overhauling their entire energy systems. Moreover, when CO_2 used in synthetic fuel production is captured from industrial processes or directly from the atmosphere, the fuel cycle can be rendered close to carbon neutral. The carbon released upon combustion is effectively offset by the carbon previously captured, forming a closed-loop system. This approach not only mitigates emissions but also provides a productive use for captured CO_2, which is essential for justifying investments in carbon capture technologies.

Economic Viability and Market Dynamics

The widespread adoption of hydrogen and synthetic fuels depends heavily on their economic competitiveness. Currently, green hydrogen is more expensive than hydrogen produced from natural gas due to the high cost of electrolysis and renewable electricity. Synthetic fuels face similar cost hurdles, as they require both green hydrogen and carbon feedstocks, along with complex synthesis technologies. Market development depends on reducing these costs through scale, innovation, and policy support. Government incentives, carbon pricing, and subsidies for clean fuel production can help bridge the gap until market forces and technological improvements drive costs down. Public-private partnerships, pilot projects, and international collaborations are playing a crucial role in demonstrating the feasibility of these fuels and attracting investment in infrastructure development.

Environmental and Lifecycle Considerations

The environmental benefits of hydrogen and synthetic fuels depend on the source of energy and feedstocks used in their production. Green hydrogen and e-fuels offer significant emissions reductions only when powered by renewable energy and supplied with captured CO_2. If fossil fuels are used at any stage, the lifecycle emissions can negate the climate benefits. Therefore, comprehensive lifecycle assessments are essential to evaluate the true sustainability of these fuels. Additionally, attention must be paid to water use in electrolysis, land use for renewable energy facilities, and potential impacts of hydrogen leaks on atmospheric chemistry. Responsible development and strict environmental standards are required to ensure that these technologies contribute positively to climate goals.

Geopolitical and Energy Security Implications

The development of hydrogen and synthetic fuels also carries important geopolitical and strategic dimensions. Countries with abundant renewable energy resources, such as sun-rich deserts or wind-prone coastlines, have the opportunity to become exporters of clean energy in the form of hydrogen or synthetic fuels. This shift could alter traditional patterns of energy trade and reduce dependence on oil and gas from politically volatile regions. At the same time, global competition for leadership in clean fuel technologies is intensifying, with regions such as the European Union, Japan, South Korea, and the Middle East launching ambitious hydrogen strategies. For many nations, investing in hydrogen and synthetic fuel capacity is not just about climate action, but also about securing a role in the future global energy economy.

Policy and Regulatory Frameworks for Advancement

Effective policy and regulation are essential for guiding the development and integration of hydrogen and synthetic fuels. Policymakers must establish clear standards for fuel quality, safety, emissions accounting, and certification. International cooperation is needed to harmonize regulations and enable cross-border trade. Incentives such as carbon pricing, tax credits, and blending mandates can stimulate demand and encourage investment. Research funding and support for demonstration projects are critical for accelerating technological maturity. National hydrogen roadmaps and synthetic fuel strategies can provide a coordinated approach that aligns stakeholders and ensures that efforts contribute meaningfully to climate targets.

6.3.3 Environmental and Economic Benefits of Carbon Reuse

Carbon reuse represents a powerful convergence of environmental responsibility and economic opportunity. By capturing and transforming carbon dioxide into valuable products, this approach addresses climate change, reduces pollution, and supports the creation of a circular carbon economy. It fosters innovation, improves resource efficiency, and opens new avenues for green growth. As the world seeks scalable and cost-effective solutions to achieve net-zero emissions, carbon reuse stands out as a pragmatic and promising pathway. With continued technological advancement, investment, and policy support, the environmental and economic benefits of carbon reuse can be fully realized, paving the way for a more sustainable and prosperous future.

Mitigating Climate Change through CO_2 Reduction

The primary environmental benefit of carbon reuse lies in its potential to reduce the amount of carbon dioxide released into the atmosphere. Climate change, driven largely by the accumulation of greenhouse gases such as CO_2, poses significant threats to ecosystems, weather patterns, and human health. Traditional carbon capture and storage (CCS) methods focus on sequestering CO_2 underground to prevent emissions, but carbon reuse adds a productive dimension by converting emissions into commercially useful products. By doing so, it contributes directly to emission reductions. When CO_2 is reused in a way that offsets the need for virgin carbon-based feedstocks or fossil fuels, it reduces the net addition of greenhouse gases to the atmosphere. This approach is particularly effective when powered by renewable energy and when CO_2 is captured from hard-to-abate sectors such as cement or steel production.

Closing the Carbon Loop and Promoting Circularity

Carbon reuse is a central component of the emerging circular carbon economy. Unlike the linear economy model, where carbon is extracted, used, and emitted, a circular carbon economy seeks to reuse and recycle carbon in a closed loop. In such systems, carbon dioxide is captured from one process and fed into another, significantly reducing the overall carbon footprint of industries. This concept promotes resource efficiency and sustainable consumption by utilizing waste carbon as a resource. For instance, carbon captured from power plants or fermentation facilities can be used to create carbonates for concrete, synthetic fuels for transportation, or feedstocks for the chemical industry. Each of these applications not only sequesters carbon temporarily or permanently but also avoids emissions that would result from using conventional fossil-derived inputs.

Improving Air Quality and Environmental Health

In addition to reducing greenhouse gas emissions, carbon reuse technologies can improve local and regional air quality. Many CO_2-intensive industries, such as power generation, refineries, and cement plants, also emit harmful pollutants like nitrogen oxides, sulfur dioxide, and particulate matter. By capturing and reusing CO_2 from these sources, the overall emissions footprint is reduced, which can lead to better air quality. This has direct health benefits for surrounding communities, lowering the incidence of respiratory diseases, cardiovascular problems, and other health issues associated with air pollution. Furthermore, when carbon reuse is combined with renewable energy and cleaner industrial processes, it amplifies the co-benefits of pollution reduction and sustainable urban development.

Stimulating Green Economic Growth

Carbon reuse also offers considerable economic benefits by fostering the growth of new industries and markets. The global push for decarbonization is stimulating demand for low-carbon and carbon-negative products, creating opportunities for businesses to innovate and expand. The development of CO_2-based products such as carbon-neutral fuels, green chemicals, and sustainable building materials is giving rise to new value chains and manufacturing sectors. As technology matures and economies of scale are achieved, carbon reuse can become a major driver of green economic growth. This includes job creation in research and development, engineering, construction, operations, and maintenance of carbon conversion facilities. By turning waste into a resource, carbon reuse enables companies to enhance profitability while contributing to climate goals.

Enhancing Resource Efficiency and Reducing Dependency on Fossil Feedstocks

Another important economic benefit of carbon reuse is its potential to reduce dependence on traditional fossil feedstocks. Many industrial processes currently rely on oil, natural gas, or coal as raw materials for producing fuels, plastics, fertilizers, and chemicals. Carbon reuse provides an alternative by offering CO_2-derived feedstocks, which can replace a portion of these fossil-based inputs. This transition reduces vulnerability to volatile fossil fuel markets and enhances energy and material security. For example, producing methanol from CO_2 and green hydrogen can substitute for fossil-derived methanol in a range of chemical applications. Likewise, using carbonates formed from CO_2 in concrete production reduces the need for energy-intensive cement components. This shift toward more sustainable inputs improves resource efficiency across multiple sectors.

Driving Innovation and Technological Advancement

The push toward carbon reuse is fueling significant technological innovation in fields such as catalysis, materials science, synthetic biology, and chemical engineering. The need to efficiently convert CO_2 into valuable products has led to the development of advanced catalysts, more efficient electrochemical cells, and engineered microorganisms that can fix CO_2 into biomass or organic acids. These technological advancements not only enhance the performance and scalability of carbon reuse systems but also have spillover effects across other clean energy and industrial applications. Furthermore, innovation is driving down the costs of carbon capture, conversion, and utilization, making these technologies more accessible to developing economies and small-scale producers.

Supporting the Decarbonization of Industrial Hubs

Carbon reuse plays a key role in supporting the decarbonization of industrial hubs and carbon-intensive clusters. Many industrial areas are co-located with power plants, chemical refineries, and manufacturing facilities that produce large volumes of CO_2. By implementing integrated carbon reuse solutions, these clusters can reduce emissions collectively and share infrastructure such as pipelines, capture units, and conversion plants. This approach improves economic feasibility and promotes regional development. It also aligns with national and international efforts to create industrial decarbonization roadmaps that are both effective and economically viable. In some regions, carbon reuse projects are becoming central to just transition plans that aim to provide new opportunities in communities historically reliant on fossil fuel industries.

Creating Value from Waste Emissions

The essence of carbon reuse lies in its ability to transform a liability into an asset. Traditionally, carbon dioxide has been treated as a waste product with associated environmental compliance costs. However, carbon reuse redefines CO_2 as a resource that can generate value. Companies that invest in carbon conversion technologies can create new revenue streams by marketing low-carbon products, selling CO_2-based fuels, or licensing proprietary technologies. This paradigm shift from waste management to resource utilization is essential in creating financial incentives for climate action. In addition, carbon reuse can support corporate sustainability goals and improve environmental, social, and governance (ESG) performance, which is increasingly important to investors and consumers.

Enabling Policy and Regulatory Support

The environmental and economic benefits of carbon reuse are being recognized by policymakers and regulators around the world. Supportive policies such as tax incentives, carbon credits, and research grants are helping to lower barriers to adoption. Regulatory frameworks that recognize and account for carbon utilization in emissions trading schemes or lifecycle assessments are also critical. These mechanisms ensure that carbon reuse projects receive appropriate recognition and rewards for their contribution to emission reductions. Furthermore, public procurement policies that favor low-carbon products can create demand and stimulate market growth. International collaboration and knowledge-sharing are enhancing best practices, accelerating deployment, and ensuring that carbon reuse becomes a mainstream element of climate strategy.

CHAPTER 7

Global CCS Initiatives and Projects

7.1 LARGE-SCALE CCS PROJECTS AROUND THE WORLD

7.1.1 Notable CCS Projects: Petra Nova, Boundary Dam, Sleipner

Petra Nova

Petra Nova is a carbon capture project designed to reduce carbon emissions from one of the boilers of a coal burning power plant in Thompsons, Texas. It is a multi-million dollar project taken up by NRG Energy and JX Nippon Oil to retrofit one of the boilers at their WA Parish Generating Station with a post-combustion carbon capture treatment system to treat a portion of the atmospheric exhaust emissions from the retrofitted boiler.

Construction and Initial Operation

The original coal-fired plant entered commercial service in 1977, and the new carbon emissions reduction system was first put into operation on January 10, 2017. The project was designed to annually capture approximately 33% of the carbon dioxide (CO_2) (or 1.6 million tonnes) emissions from the plant's boiler #8. The carbon dioxide gas was captured at 99% purity, and is then compressed and piped about 82 miles to the West Ranch Oil Field, where it is used for enhanced oil recovery. The oil field had previously been producing 300 barrels of oil per day. With the new injection of high pressure carbon dioxide into the field, the oil production of the field was increased by a factor of 50 to 15,000 barrels per day. This project was expected to run for at least another 20 years. To satisfy the Clean Coal Power Initiative requirements, the Texas Bureau of Economic Geology put a monitoring plan in place to keep track of the injection and movement of the CO_2 beneath the surface and in the rock structures at West Ranch. On May 1, 2020, NRG shut down Petra Nova, citing low oil prices during the COVID-19 pandemic. Due primarily to outages of the host natural gas cogeneration unit, the project missed its carbon sequestration goal by 17% over its first three years of operation.

Restarting

On September 14, 2022, Japan's Eneos Holdings Inc announced plans to acquire full ownership of Petra Nova Parish Holdings. JX Nippon Oil & Gas Exploration,

Eneos' subsidiary, purchased the remaining 50% stake from NRG Energy Inc for $3.6 million. Eneos' strategy was to gain expertise in carbon capture, utilization, and storage (CCUS) technology through this move, and they announced plans to restart the unit in the second quarter of 2023. On September 13, 2023, JX Nippon Oil & Gas Exploration Corporation announced the resumption of operations beginning on September 5, 2023.

Fig: Petra Nova Shown on the Right

Technology

The Petra Nova carbon emissions reduction system uses an amine-based absorption system, or the KM CDR Process (Kansai Mitsubishi Carbon Dioxide Recovery). This process was developed by Mitsubishi Heavy Industries and Kansai Electric Power and utilizes a high performing proprietary solvent called KS-1. The CO_2 is removed from the exhaust gas through a basic absorber-stripper system. The gaseous CO_2 is then compressed to a supercritical liquid. The CO_2 leaving the carbon capture plant is over 99% pure and is sent 82 miles through 12-inch diameter pipes to their end location of the West Ranch oil field, where it is used for enhanced oil recovery. The carbon dioxide from the Petra Nova Initiative will eventually end up in sandstone in the Frio Formation of the West Ranch oil field. It will remain about 5,000 feet underground and cover more than 4,000 acres of ground area.

Economics

The Petra Nova carbon emissions reduction system cost approximately $1 billion to install, and received a nearly $190 million grant from the U.S. Government under the Clean Coal Initiative, as well as a $250 million loan from the Japanese government. The increased oil recovery in the adjacent oil field was expected to result in a net savings. However, when the project was first proposed oil prices were very high (at $100 per barrel) and it was assumed that they would not drop. As of 2017, the current oil price was about $50 per barrel and therefore a net loss has resulted regarding the oil production at the field. On May 1, 2020, NRG shut down Petra Nova, citing the low cost of oil as a result of the COVID-19 pandemic.

Boundary Dam

Boundary Dam is a concrete arch gravity-type hydroelectric dam, finished in 1967, on the Pend Oreille River, in the U.S. state of Washington. The dam is located in the northeast corner of Washington state. It is operated by Seattle City Light and makes up a significant portion of the City of Seattle's energy portfolio. On average, it provides upwards of 46% of the power generated by Seattle City Light. Boundary Powerhouse, located adjacent to the dam, is completely built inside of the rock that makes up the left abutment of the dam itself. It has a nameplate capacity of just over 1 gigawatt of generation. The component of the hydroelectric project were listed on the National Register of Historic Places in 2018.

Fig: Boundary Dam

Sleipner Gas Field

The Sleipner gas field is a natural gas field in the block 15/9 of the North Sea, about 250 kilometres (160 mi) west of Stavanger, Norway. Two parts of the field are in production, Sleipner West (proven in 1974), and Sleipner East (1981). The field produces natural gas and light oil condensates from sandstone structures about 2,500 metres (8,200 ft) below sea level. It is operated by Equinor. The field is named after the eight-legged steed Sleipnir of Odin a widely revered god in Norse mythology.

Reserves and Production

As of the end of 2005, the estimated recoverable reserves for the Sleipner West and East fields were 51.6 billion cubic meters of natural gas, 4.4 million tonnes (4.9 million short tons) of natural gas liquids, and 3.9 million cubic meters of condensates. Daily production of the field in 2008 was 300 thousand bbl (48 thousand m3) oil equivalents per day, 36 million cubic meters of natural gas per day, and 14,000 cubic meters of condensate per day. In an updated 2017 report, the Norway Petroleum Directorate estimates 2.72 million cubic meters of oil, 11.72 billion cubic meters of natural gas, 0.67 million tons of natural gas liquids, and 0.07 million cubic meters of condensates remain in the reserves. Sleipner field consists of four platforms. The field is planted with 18 production wells. The Sleipner A platform is located on the Sleipner East and the Sleipner B platform is located on the Sleipner West. Sleipner B is operated remotely from the Sleipner A via an umbilical cable. The Sleipner T carbon dioxide treatment platform is linked physically to the Sleipner A platform by a bridge and to the Sleipner B wellhead platform by 12.5-kilometre (7.8 mi) carbon dioxide flow line. The Sleipner Riser platform, serving the Langeled and Zeepipe pipelines, is located on the Sleipner East field.

Carbon Capture and Storage Project

The Sleipner Vest (West) field is used as a facility for carbon capture and storage (CCS). It is the world's first offshore CCS plant, operative since September 15, 1996. The project, in the initial year, proved insecure due to sinking top sand. However, after a re-perforation and an installation of a gravel layer in August 1997, CCS operations were secure. Equinor reported that as of 2018, one million tonnes of CO_2 have been transported and injected into the formation yearly since 1996. Equinor later stated that due to a problem with monitoring equipment, it had over-reported the amount of CO_2 sequestered from 2017 to 2019 and released new figures that were around 30% less. The project summary reports a capacity of up to 600 billion tonnes (≈660 billion tons).

The Sleipner West field has up to 9% CO_2 concentration; Norway only allows 2.5% CO_2 before imposing production export quality penalties, which may have been NOK 1 million/day (~$120,000US/ day). Operating costs are US$17 / ton of CO_2 injected, however, the company does not pay Norway's carbon tax of 1991 and receives carbon credit in the EU's emissions trading system. Before the carbon tax, industries released poor quality CO_2 into the atmosphere. In a business-as-usual scenario, Norway's emissions would have had a total increase of 3% over 20 years if not for the CCS experiment. Carbon dioxide is treated on the Sleipner T treatment platform. After that carbon dioxide is transported to the Sleipner A platform where it is injected into the Utsira formation through a dedicated well c. 1000 meters under the seabed. Using time-lapse gravity and seismic methods, the pioneering Sleipner carbon capture project confirmed the technological viability of injecting and measuring CO_2 in an offshore reservoir, as well as the effectiveness of mitigating emissions through stable storage. To avoid possible leakages that can result in health hazards and environmental destruction, above the Utsira Formation injection site lies 30 seafloor gravity stations for monitoring under the title, Saline Aquifer CO_2 Storage. These sites monitor microseismic activity along with gravitational forces and depth metrics. Seafloor height, natural gas production, and tidal shifts determine the gravity measured.

Explicitly regulated under Norway's petroleum law in December 2014 and in line with the EU's 2009/31/EC directive, monitoring objectives focus on assessing gas movement, shell stability, and the effectiveness of remedy scenarios in case of leakage. From 2002 to 2005, measurements identified vertical changes in established metric boundaries, most likely attributed to erosion and marine life. Onsite geochemical and reservoir simulations reveal a main buildup of CO_2 under the formation's cap seal. However, when the injections are eventually decommissioned, simulations show accumulation proximate to the cap seal in clay layers saturated with sand, which will result in solubility trapping. This solubility trapping, caused by the multiple layers of clay and sand, prevents CO_2 from rising beyond and will ultimately turn to mineral trapping in the substrate. Furthermore, groundwater flow facilitates better distribution of gases and depressurization, lowering the risk of leakage. The composition reaction of the mixture of clay, sand, and carbon is the determining factor of long-term stability in the Sleipner CCS project. As of 2007, measurements from the gravity stations revealed that the injection of CO_2 into the Utsira Formation has not resulted in any noticeable seismic activity and that there have been no carbon dioxide leakages in the past 10 years.

Natural gas pipelines' operator Gassco had proposed to build a 240-kilometre (150 mi) carbon dioxide pipeline from Kårstø to transport carbon dioxide from the now decommissioned Kårstø power station. While injection pipelines do not succumb to rusting when transporting CO_2, transport pipelines experience low temperatures and high pressures, resulting in dew formation, and subsequently, rust.

Miocene Utsira Formation

The Miocene Utsira Formation is a large aquifer with a stable, layered clay seal. Distributed through multiple phases as a result of sea-level variations caused by glacial events in the Pliocene period, deposits date back to the late Miocene/ early Pliocene to early Pleistocene times, determined by palynology. Upper Pliocene deltaic sand deposits blanket the formation with the highest top sands located roughly 150 meters below sea level. Measured with 3D seismic data, the Utsira sandstone lies underneath 800–1000 meters of sediment under the sea with a maximum thickness of over 300 meters. The Utsira stretches 450 kilometers north to south and 90 kilometers east to west. In the north and south lie deep sand systems, while in the middle region slimmer deposits cover the seafloor. The Tampen area, located in the most northern region, contains lean deposits of glauconitic sand.

7.1.2 Lessons Learned from Ongoing CCS Projects

The experiences of Petra Nova, Boundary Dam, and Sleipner serve as important case studies in the global development of carbon capture and storage. They reveal both the promise and complexity of implementing CCS at scale. While each project is shaped by its unique technical and economic context, the common insights gained are invaluable for designing more efficient, cost-effective, and socially acceptable CCS systems. As climate goals become increasingly urgent, the lessons from these early projects will help inform the next generation of CCS initiatives, ensuring that they are more resilient, integrated, and impactful in reducing global greenhouse gas emissions.

Petra Nova: Advancing Post-Combustion Capture in Power Generation

The Petra Nova project in Texas represented one of the most ambitious efforts to retrofit a coal-fired power plant with carbon capture technology. Operational from 2017 to 2020, the project aimed to capture over 1 million tons of CO_2 annually from a 240-megawatt slipstream of the W.A. Parish coal plant using post-combustion capture technology. The captured CO_2 was then transported via pipeline and injected into the West Ranch oil field for enhanced oil recovery (EOR). One of the key lessons from Petra Nova is the importance of economic viability in the success of CCS projects. Although the system worked as designed and demonstrated that large-scale carbon capture at a power plant is technically feasible, the project's economic foundation was strongly tied to the price of oil. When oil prices fell significantly, the financial justification for the project weakened, ultimately leading to its suspension in 2020. This outcome underscores the need for stable and diversified revenue streams, possibly through carbon pricing or regulatory support, to ensure the longevity of CCS investments. Another important takeaway from Petra Nova is the role of integrated project design and operations. The use of Mitsubishi Heavy Industries' amine-based capture system demonstrated high efficiency and reliability during the operating period. However, the integration between the capture facility, the power plant, and the EOR infrastructure required complex coordination. Lessons from Petra Nova suggest that early and detailed planning across all components of a CCS chain is essential to minimize operational disruptions and optimize performance. Furthermore, the project highlighted the significance of public-private partnerships and government support in the form of tax credits and grants to enable large-scale deployment of CCS technologies in the absence of stringent carbon regulations.

Boundary Dam: Pioneering CCS in Coal Power and Operational Learning

Boundary Dam in Saskatchewan, Canada, was the first full-scale CCS facility integrated into a commercial coal-fired power plant. Operational since 2014, the project retrofitted Unit 3 of the Boundary Dam Power Station with a post-combustion carbon capture system capable of capturing up to 1 million tons of CO_2 annually. The captured CO_2 is transported to nearby oil fields for EOR, with excess CO_2 stored in deep saline aquifers. One of the major lessons from Boundary Dam is the value of transparency and knowledge sharing. SaskPower, the project operator, has actively shared data and operational insights with the global CCS community, helping advance understanding of capture technology performance, cost, and reliability.

Boundary Dam has faced several technical challenges since its commissioning, including maintenance issues with the CO_2 compressor and downtime that affected capture rates. These experiences emphasize the need for robust maintenance strategies, quality control in equipment procurement, and contingency planning for complex industrial systems. Importantly, the project has evolved over time, with improvements in reliability and cost reductions that make future retrofits more viable. The project also demonstrated the necessity of pilot testing and phased implementation, as real-world operational conditions can differ significantly from lab or pilot-scale projections.

From an environmental perspective, Boundary Dam has contributed significantly to reducing CO_2 emissions while allowing the continued use of coal power in a lower-carbon framework. The experience gained has informed SaskPower's decision-making regarding the future of coal generation and the role of CCS in the province's broader decarbonization strategy. Additionally, public engagement efforts around Boundary Dam revealed the importance of educating stakeholders about the benefits and safety of CCS, particularly in regions where fossil fuels remain economically and socially important.

Sleipner: Setting the Benchmark for Offshore CO_2 Storage

Sleipner, operated by Equinor (formerly Statoil), holds the distinction of being the world's first commercial-scale CCS project. Since 1996, it has captured and stored approximately one million tons of CO_2 annually from the natural gas processing operations at the Sleipner West field in the North Sea. Unlike Petra Nova and Boundary Dam, which focus on capturing emissions from coal combustion, Sleipner captures CO_2 directly from produced natural gas, which must be stripped of CO_2 to meet pipeline specifications. The CO_2 is then injected into a deep saline formation known as the Utsira Sand, a large and well-characterized geological storage site. Sleipner provides several key lessons for the global CCS community. First, it has demonstrated the long-term technical feasibility and safety of offshore CO_2 storage. Decades of monitoring using seismic imaging, pressure measurements, and modeling have shown that the injected CO_2 remains securely trapped within the geological formation. This data is invaluable for building public confidence and regulatory frameworks for CCS deployment. The project illustrates the importance of selecting appropriate geological sites for storage, with emphasis on porosity, cap rock integrity, and pressure management.

Economically, Sleipner highlights the role of regulatory drivers in enabling CCS. The project was developed largely in response to Norway's carbon tax, which made venting CO_2 more expensive than storing it. This policy incentive aligned economic and environmental objectives, demonstrating that clear and consistent carbon pricing can catalyze investment in emission reduction technologies. Moreover, the simplicity and self-contained nature of Sleipner's capture and storage system contributed to its operational stability and cost-effectiveness, particularly in comparison to more complex capture-from-combustion processes.

Sleipner's success has also contributed to the establishment of best practices in CO_2 monitoring, reporting, and verification (MRV). It set an early standard for transparent documentation of storage performance, helping shape international guidelines and regulatory expectations. The project underscores the importance of long-term monitoring not only for safety and compliance but also for building scientific understanding and public trust in CCS technologies.

Cross-Cutting Lessons and Implications for Future CCS Deployment

Together, Petra Nova, Boundary Dam, and Sleipner offer a comprehensive view of the opportunities and challenges facing CCS implementation. One of the most important cross-cutting lessons is that successful CCS projects require alignment between technological readiness, economic incentives, and supportive policy frameworks. Whether through tax credits, carbon pricing, or public funding, financial support is essential to overcome the high upfront capital costs and operational risks associated with CCS technologies. Another shared lesson is the necessity of public engagement and transparency. All three projects underscore the importance of building stakeholder trust through clear communication, reliable monitoring, and evidence-based assessments of safety and effectiveness.

Technical reliability is also a critical theme. While the core technologies for capture and storage are proven, real-world deployment often reveals unexpected challenges that require adaptive management, ongoing investment in R&D, and workforce training. Finally, the diversity of applications represented by these projects—from coal power and natural gas processing to offshore storage—illustrates the flexibility of CCS across various sectors and geographies. This adaptability is key to making CCS a foundational element of global decarbonization strategies. Moving forward, future projects will benefit from the operational data, lessons learned, and public documentation provided by these pioneering efforts.

7.2 REGIONAL CCS DEVELOPMENT: NORTH AMERICA, EUROPE, ASIA

7.2.1 U.S. CCS Programs: The Role of DOE and National Labs

Regional CCS development in North America, Europe, and Asia reflects the diverse challenges and opportunities associated with deploying carbon capture technologies at scale. North America's experience highlights the importance of infrastructure and policy incentives, Europe's leadership illustrates the role of regulatory innovation and climate ambition, and Asia's evolving strategies demonstrate the balancing act between industrial growth and environmental responsibility. As each region continues to advance its CCS agenda, the cumulative experience and lessons learned will shape a more coordinated and effective global response to climate change. Building on regional strengths and fostering international partnerships will be essential to realizing the full potential of CCS in achieving a sustainable and low-carbon future.

North America: Leadership in CCS Infrastructure and Innovation

North America, particularly the United States and Canada, has been at the forefront of CCS development for decades. The region benefits from a mature oil and gas industry, well-characterized geological formations for CO_2 storage, and an extensive network of pipelines that facilitate the transportation of carbon dioxide. The presence of these infrastructure assets has supported large-scale CCS projects such as Petra Nova in Texas, Boundary Dam in Saskatchewan, and several enhanced oil recovery operations that utilize captured CO_2 to boost petroleum production.

In the United States, CCS deployment has been significantly supported by federal incentives, especially the 45Q tax credit, which provides financial compensation per ton of CO_2 captured and either stored permanently or used in EOR. This policy has spurred renewed interest in CCS, especially in power generation, cement, steel, and hydrogen production. The Infrastructure Investment and Jobs Act and the Inflation Reduction Act further expanded funding for CCS research and development, including support for carbon capture hubs and demonstration projects. These hubs are expected to enable shared infrastructure and reduce the cost of deployment across industries. Canada has also played a leading role, particularly in integrating CCS with oil sands and heavy industry. Projects like the Quest CCS facility in Alberta and the Alberta Carbon Trunk Line demonstrate how CCS can be applied in complex industrial environments. Canada's approach also emphasizes collaboration between government, industry, and research institutions. The country's federal and provincial carbon pricing schemes provide additional motivation for companies to reduce emissions using CCS technologies. One of the defining features of North American CCS is the combination of public support, policy incentives, and private investment that have made it a testing ground for scaling CCS in diverse sectors.

Europe: Regulatory Innovation and Climate Ambition Driving CCS

Europe has approached CCS from a strong climate policy standpoint, integrating it as a key element of its broader decarbonization strategy under the European Green Deal and Fit for 55 package. Although CCS deployment in Europe has faced delays and public resistance in some regions, the continent is experiencing renewed momentum, particularly in countries like Norway, the Netherlands, and the United Kingdom. Europe's emphasis on stringent emissions targets and its carbon pricing mechanisms under the EU Emissions Trading System (EU ETS) have helped create a favorable environment for CCS investment.

Norway has been a pioneer in CCS through its long-running Sleipner and Snøhvit offshore storage projects, which demonstrate the technical feasibility and safety of geological sequestration beneath the North Sea. These projects have also contributed to global CCS knowledge through extensive monitoring and reporting practices. Building on this legacy, Norway is now leading the Northern Lights project, part of the Longship initiative, which aims to develop a full-scale CCS value chain involving capture, transport, and permanent storage of CO_2 from multiple industrial sources across Europe. This cross-border collaboration reflects Europe's commitment to regional integration and shared infrastructure.

The United Kingdom has also made significant strides, establishing policies and funding mechanisms to support industrial decarbonization through CCS. The UK's Industrial Decarbonisation Strategy outlines plans for industrial clusters equipped with shared capture and storage facilities. Projects like Net Zero Teesside and the East Coast Cluster are among the most ambitious in Europe, aiming to capture millions of tons of CO_2 annually from power plants, hydrogen production, and chemical facilities. The UK government has committed substantial funding and policy support to accelerate deployment, including contracts for difference mechanisms to ensure revenue certainty for CCS operators.

In the Netherlands, CCS is being prioritized to meet short- and medium-term climate goals. The Porthos project in the Port of Rotterdam aims to store CO_2 beneath the North Sea, leveraging the region's dense industrial activity and existing infrastructure. European CCS initiatives emphasize the integration of renewable energy and hydrogen production, with CCS acting as a bridging technology to ensure emissions reductions while low-carbon energy systems mature. Europe's approach showcases how regulatory clarity, public funding, and international collaboration can help overcome the financial and social challenges of CCS deployment.

Asia: Rapid Industrialization Meets Emerging CCS Strategies

Asia presents a complex and diverse landscape for CCS development, shaped by the region's rapid industrial growth, high energy demands, and varying levels of economic development. Countries such as China, Japan, South Korea, and India are recognizing the importance of CCS in decarbonizing their energy and industrial sectors, particularly in coal-dependent economies. However, the scale of emissions and the cost of CCS implementation pose significant challenges.

China, the world's largest emitter of CO_2, has made CCS part of its long-term strategy to achieve carbon neutrality by 2060. The country has launched several pilot projects and demonstration plants, including the Ordos CCS project and efforts within the coal-to-chemicals industry. The development of carbon capture hubs in industrial regions such as Yanchang and Xinjiang aims to aggregate emissions from multiple sources and direct them to shared storage sites. China's approach emphasizes domestic technology development, integration with EOR, and alignment with its national carbon trading system. While progress has been steady, greater policy coordination, financial support, and public-private partnerships are necessary to scale CCS across China's vast industrial base.

Japan and South Korea, both highly industrialized and energy-import-dependent nations, have also prioritized CCS in their climate action plans. Japan's Tomakomai CCS demonstration project successfully injected and monitored CO_2 offshore and is being followed by broader initiatives under the Green Growth Strategy. Japan envisions CCS as crucial to its blue hydrogen ambitions, with captured CO_2 from fossil-based hydrogen production being stored to achieve carbon neutrality. Similarly, South Korea's CCS Roadmap focuses on scaling up capture and storage capacity to reduce industrial emissions and support the production of clean hydrogen and ammonia. These countries are also investing in international CCS collaborations, including cross-border storage opportunities in Southeast Asia and Oceania.

India's CCS strategy is still in early stages, but interest is growing, especially in the context of decarbonizing cement, steel, and thermal power generation. The availability of significant geological storage capacity and technical expertise from the oil and gas sector provides a foundation for future CCS development. However, financial constraints, limited regulatory frameworks, and competing development priorities present hurdles that must be addressed through international cooperation, technology transfer, and targeted policy interventions.

Common Challenges and Opportunities across Regions

Despite regional differences, several common themes emerge in the development of CCS in North America, Europe, and Asia. High capital costs, uncertainty in long-term revenue streams, and the need for supportive policy environments are universal concerns. Technical feasibility has been largely demonstrated, but commercial scalability requires more robust incentives, standardized regulations, and greater public acceptance. The establishment of carbon capture hubs and shared transport and storage infrastructure is a promising strategy to reduce costs and aggregate emissions reductions across industries. Another shared opportunity lies in the integration of CCS with other clean technologies, such as hydrogen production, bioenergy, and renewable power. This synergy can multiply the climate benefits of CCS while unlocking new economic opportunities. International collaboration, knowledge sharing, and harmonization of standards can accelerate deployment and ensure that CCS plays a meaningful role in the global transition to net-zero emissions.

7.2.2 Europe's CCS Strategy and the Role of EU ETS

Europe's commitment to achieving climate neutrality by 2050 has brought Carbon Capture and Storage (CCS) into sharp focus as a crucial technology in its transition to a decarbonized economy. While the continent has made significant advances in renewable energy, electrification, and energy efficiency, CCS is increasingly recognized as indispensable for reducing emissions from hard-to-abate sectors such as cement, steel, chemical manufacturing, and waste incineration. The European Union's (EU) strategic approach to CCS reflects both climate urgency and economic pragmatism, combining regulatory frameworks, targeted funding, and infrastructure development. At the heart of this strategy lies the European Union Emissions Trading System (EU ETS), which serves as the primary market-based instrument for carbon pricing and incentivizing emissions reduction technologies, including CCS.

Europe's CCS strategy, anchored by the EU ETS, reflects a comprehensive approach to decarbonization that balances environmental integrity with economic feasibility. The rising carbon price, combined with targeted funding and supportive regulation, is gradually creating a viable market for CCS. While challenges remain, including high costs, public skepticism, and regulatory complexity, Europe has established a strong policy foundation and a pipeline of projects that signal increasing momentum. By integrating CCS into industrial clusters, cross-border networks, and low-carbon energy systems, the EU is not only advancing its own climate goals but also setting global benchmarks for responsible and scalable CCS deployment.

Policy Foundation for CCS in Europe

Europe's CCS strategy is driven by the European Green Deal, which sets out a roadmap for transforming the EU into a climate-neutral region by mid-century. As part of this framework, the EU has adopted increasingly stringent climate targets, including a 55% reduction in greenhouse gas emissions by 2030 compared to 1990 levels. These targets are legally binding under the European Climate Law and are reinforced by sector-specific legislation and initiatives such as the Fit for 55 package. Within this broader policy landscape, CCS is identified as a critical enabler of deep decarbonization, particularly for sectors that lack viable alternatives to fossil fuel-based processes.

Several EU directives and initiatives explicitly support the development and deployment of CCS. The Directive on the Geological Storage of Carbon Dioxide provides a legal framework for the safe and environmentally sound storage of CO_2 in geological formations, setting standards for site selection, monitoring, and liability. In addition, the EU Innovation Fund and Connecting Europe Facility allocate funding for large-scale CCS projects, especially those demonstrating advanced technologies and cross-border infrastructure. These policy tools complement national initiatives in member states like Norway, the Netherlands, and the United Kingdom, each of which has developed ambitious national strategies to integrate CCS into industrial and energy planning.

The EU ETS in Promoting CCS Deployment

The European Union Emissions Trading System is the cornerstone of the EU's climate policy and serves as the most influential driver of CCS investment. Launched in 2005, the EU ETS is the world's largest carbon market, covering more than 10,000 installations in the power, industrial, and aviation sectors. It operates on a cap-and-trade principle, whereby a finite number of emission allowances are issued, and companies can buy or sell these permits based on their emissions levels. The declining cap on allowances ensures that emissions decrease over time, while the trading mechanism introduces market efficiency and cost-effectiveness.

CCS plays a specific and strategic role within the EU ETS framework. According to the EU ETS Directive, installations that capture, transport, and store CO_2 in compliance with the CCS Directive are considered to have zero emissions for the stored amount of CO_2. This provision gives CCS installations a direct financial benefit by reducing their obligation to surrender allowances, effectively lowering the cost of compliance. As the carbon price under the EU ETS has risen significantly in recent years, reaching over €90 per ton in 2022, the economic case for investing in CCS has become increasingly attractive for companies facing high emissions costs.

In addition to reducing compliance costs, the EU ETS indirectly supports CCS by creating a revenue stream for governments through auctioning allowances. These revenues can be reinvested into climate-related initiatives, including CCS research, infrastructure, and deployment. Many member states have used ETS revenues to co-finance CCS pilot projects and develop regional carbon capture hubs. Furthermore, the growing scarcity and value of allowances increase pressure on carbon-intensive industries to adopt technologies like CCS to remain competitive and meet regulatory obligations.

Flagship CCS Projects and Cross-Border Infrastructure in Europe

The implementation of CCS in Europe is taking shape through a number of high-profile projects that aim to demonstrate feasibility and create economies of scale.

One of the most significant initiatives is Norway's Longship project, which includes the Northern Lights transport and storage network in the North Sea. This project enables the permanent offshore storage of CO_2 from multiple industrial sources, both domestic and international. The Northern Lights initiative is particularly notable for its open-access model, allowing emitters across Europe to utilize the infrastructure, thereby reducing entry barriers for smaller emitters.

In the Netherlands, the Porthos project seeks to establish a similar infrastructure in the Port of Rotterdam. It plans to capture CO_2 from local industrial facilities and transport it via pipeline for offshore storage beneath the North Sea. The project has secured EU funding and is backed by a consortium of government and private stakeholders. Porthos exemplifies the EU's strategy of using CCS to decarbonize industrial clusters and maximize infrastructure efficiency.

The United Kingdom, though no longer part of the EU, continues to influence the broader European CCS landscape through projects such as the East Coast Cluster and HyNet. These initiatives focus on decarbonizing industrial regions using integrated CCS and hydrogen production systems. Their development illustrates how shared transport and storage infrastructure can accelerate CCS deployment and reduce costs per ton of CO_2 stored.

Challenges and Social Dimensions of CCS in Europe

Despite policy momentum and promising projects, Europe's CCS strategy faces several challenges. High capital and operational costs remain a barrier, particularly for first-of-a-kind projects. While the EU ETS provides a carbon price signal, the price alone may not be sufficient to guarantee bankable returns on CCS investments without additional support mechanisms. The EU's Innovation Fund aims to bridge this gap, but the scale of funding remains limited relative to the vast capital required for continent-wide CCS deployment.

Public perception and societal acceptance also pose significant hurdles. CCS has encountered opposition in some European countries due to safety concerns, lack of awareness, and association with prolonging fossil fuel use. Past project cancellations in countries like Germany and the Netherlands demonstrate how local resistance can derail even well-planned initiatives. As a result, effective communication, public engagement, and transparent risk assessments are essential components of Europe's CCS strategy. Building public trust requires not only technical assurances but also a compelling narrative that links CCS to climate protection, job creation, and regional economic resilience. Another concern is regulatory fragmentation across member states. While the EU provides overarching frameworks, national-level differences in permitting, liability rules, and financial incentives can slow down project development and increase transaction costs. Harmonizing regulations and streamlining permitting processes would significantly enhance CCS deployment, particularly for cross-border projects that depend on coordination between multiple jurisdictions.

Integrating CCS with Europe's Broader Climate Agenda

CCS in Europe is not a standalone solution but a key enabler within a broader decarbonization portfolio. The EU's strategy emphasizes the integration of CCS with clean hydrogen production, known as blue hydrogen, which relies on methane reforming with carbon capture. This approach allows Europe to reduce emissions from gas use while scaling up green hydrogen infrastructure powered by renewables. CCS is also integral to achieving negative emissions through Bioenergy with Carbon Capture and Storage (BECCS), which combines biomass combustion or gasification with carbon capture to remove CO_2 from the atmosphere.

Moreover, CCS supports the circular carbon economy by enabling carbon reuse and synthetic fuel production. Captured CO_2 can be utilized to manufacture fuels, chemicals, and building materials, thus creating new market opportunities and fostering innovation. These applications reinforce the EU's commitment to industrial decarbonization and green growth.

7.2.3 CCS Adoption in Asia and the Global South

The adoption of CCS in Asia and the Global South is essential for achieving global climate goals, particularly given the scale of emissions and projected economic growth in these regions. While challenges persist in terms of cost, policy, and infrastructure, momentum is building through pilot projects, international partnerships, and increasing recognition of CCS in national climate strategies. The future of CCS in these regions will depend on targeted support, inclusive policies, and sustained investment in research, demonstration, and deployment. As the world moves toward a decarbonized future, ensuring that CCS is accessible, affordable, and effectively implemented in the Global South is not only an environmental imperative but also a matter of global equity and shared responsibility.

Growing Importance of CCS in Asia

Asia is home to some of the world's largest carbon emitters, including China and India, and is characterized by high levels of industrial activity and dependence on fossil fuels. As such, CCS offers a practical solution to decarbonize hard-to-abate sectors including coal-fired power, cement, steel, and petrochemicals. China has recognized CCS as a strategic component of its climate policy, particularly in its effort to achieve peak carbon emissions before 2030 and carbon neutrality by 2060. The Chinese government has included CCS in its Five-Year Plans and has initiated multiple pilot and demonstration projects. Notable among them are the

Sinopec Qilu Petrochemical CCS project and the Shenhua coal-to-liquid project, which demonstrate the integration of CCS into energy and chemical production systems. China's CCS strategy is focused not only on emission reduction but also on technology development and industrial competitiveness. Research institutions and state-owned enterprises are actively investing in capture technologies, storage site assessments, and CO_2 utilization. However, the challenge remains in scaling these efforts from pilot phases to commercial deployment. Financial viability is a key barrier, as the costs of CCS remain high without significant market incentives or carbon pricing mechanisms. While China has launched a national carbon trading system, its current coverage and pricing levels are not yet robust enough to drive widespread CCS adoption. Strengthening regulatory frameworks and offering targeted subsidies or tax incentives could help accelerate CCS development in the country.

India's Path toward CCS Integration

India presents a different but equally pressing case for CCS adoption. As a rapidly developing country with a large population and growing energy needs, India faces the complex task of balancing economic growth with environmental sustainability. The country is heavily reliant on coal for electricity generation and has a significant carbon footprint from industries such as cement and steel. Despite this, CCS has not yet gained substantial policy attention or investment in India compared to other clean energy technologies like solar and wind. There is, however, increasing interest from academic institutions, industry players, and government agencies in exploring CCS as part of India's long-term decarbonization strategy. Studies by the Indian government's Department of Science and Technology and the Council on Energy, Environment and Water have emphasized the importance of building technical capacity, identifying suitable storage sites, and creating an enabling policy environment. India has considerable potential for geological storage, particularly in its sedimentary basins, but detailed site characterization and regulatory oversight are still in early stages. CCS in India will likely evolve through public-private partnerships and international cooperation, especially in the form of technology transfer, climate finance, and pilot projects supported by multilateral institutions.

Southeast Asia's Emerging Interest in CCS

Southeast Asia, with its mix of developing economies and reliance on fossil fuels, is increasingly looking at CCS as a viable solution for long-term emissions reduction. Countries like Indonesia, Malaysia, and Vietnam have begun to integrate CCS into national energy and climate plans. Indonesia, for example, has included CCS in its Enhanced Nationally Determined Contributions (NDCs) and is collaborating with

Japanese and Australian partners to develop pilot projects in natural gas processing and coal-based power generation.

Malaysia has also shown momentum in CCS development, particularly through Petronas, its national oil company, which is exploring offshore storage and blue hydrogen production. The region's proximity to offshore sedimentary basins offers promising opportunities for CO_2 storage, but challenges include the need for capacity building, regulatory clarity, and substantial capital investment. Regional collaboration through organizations such as the ASEAN Centre for Energy can play a key role in harmonizing policies, sharing technical knowledge, and attracting foreign investment.

CCS in the Global South

Beyond Asia, the Global South includes countries in Latin America, Africa, and the Middle East that are increasingly engaging with CCS as a component of their climate and energy strategies. In the Middle East, countries such as the United Arab Emirates and Saudi Arabia are investing heavily in CCS, driven by a desire to decarbonize their oil and gas sectors while maintaining energy export revenues. The UAE's Al Reyadah project and Saudi Arabia's Uthmaniyah CO_2-EOR initiative are among the most advanced CCS operations in the region and serve as benchmarks for other nations.

Africa, while contributing minimally to global emissions, faces unique challenges in CCS deployment. The continent's priorities are largely centered on energy access, poverty alleviation, and sustainable development. Nevertheless, countries like South Africa have acknowledged CCS in their national climate policies. South Africa's high reliance on coal and its significant industrial base make it a strong candidate for CCS, and pilot projects such as the South African Centre for Carbon Capture and Storage have been initiated to explore feasibility. However, widespread CCS adoption will depend on international support in the form of financing, technology transfer, and capacity development.

Latin America also shows emerging CCS activity, particularly in Brazil, which has begun evaluating CCS options in the context of bioenergy and oil and gas. Petrobras has initiated projects involving CO_2 injection in pre-salt oil fields, and there is growing interest in combining CCS with bioenergy (BECCS) to generate negative emissions. Argentina and Colombia are similarly evaluating the potential for CCS in natural gas and industrial applications. These efforts are supported by regional cooperation initiatives and partnerships with institutions such as the Inter-American Development Bank and international research networks.

Barriers and Opportunities in CCS Expansion

Despite promising developments, several barriers hinder the widespread adoption of CCS in Asia and the Global South. Chief among these is the high cost of capture, transport, and storage infrastructure, which is especially burdensome in developing economies with limited fiscal space. The lack of mature carbon markets, weak regulatory frameworks, and limited technical expertise further constrain progress. Additionally, public awareness of CCS remains low, and there is often skepticism about its safety, effectiveness, and long-term benefits.

However, these challenges also present opportunities for innovation and collaboration. Multilateral climate finance mechanisms such as the Green Climate Fund, the Global Environment Facility, and the Climate Investment Funds can play a critical role in de-risking CCS investments and supporting capacity building. Technology partnerships between developed and developing countries can accelerate learning and reduce costs through shared infrastructure and joint ventures. Moreover, the development of regional CCS hubs, modeled on those in Europe and North America, can enable economies of scale and collective learning in the Global South.

7.3 INTERNATIONAL COOPERATION AND GLOBAL CCS NETWORKS

7.3.1 Collaborative Research and Development Initiatives

The global challenge of climate change necessitates equally global solutions, and the development of Carbon Capture and Storage technologies is no exception. International cooperation and collaborative R&D initiatives are essential for accelerating CCS deployment, reducing costs, ensuring safety, and enabling equitable access to this critical technology. Through global CCS networks, countries can share knowledge, align regulations, pool resources, and build the infrastructure necessary to scale CCS beyond isolated projects. These efforts not only contribute to emissions reduction but also foster technological innovation, capacity development, and climate equity. As the world approaches critical climate deadlines, the role of collaboration in CCS becomes ever more vital, offering a model for how collective action can turn complex challenges into shared progress.

International Cooperation in CCS Development

The technological and financial demands of CCS require unprecedented levels of collaboration. Unlike renewable energy systems that can often be deployed at smaller, modular scales, CCS involves complex infrastructure across capture, transport, and storage stages, each of which demands site-specific knowledge, substantial investment, and technical expertise. This makes global cooperation critical, especially for early-stage research and pilot demonstrations that help de-risk large-scale deployment. Furthermore, many nations possess only part of the CCS value chain, such as CO_2-emitting industries but not suitable storage sites.

International networks enable these countries to participate in shared solutions, such as transboundary CO_2 transport and offshore storage. In the context of international climate commitments like the Paris Agreement, cooperative CCS R&D also supports technology transfer and equitable participation by developing nations. Without inclusive participation, the world risks exacerbating technological and economic divides. Collaborative CCS initiatives can help ensure that all countries, regardless of income level, can contribute to and benefit from climate mitigation technologies.

Major International CCS Networks and Organizations

Several global and regional organizations play leading roles in coordinating CCS collaboration. One of the foremost entities is the Carbon Sequestration Leadership Forum (CSLF), established in 2003 to facilitate the development of cost-effective CCS technologies through multilateral cooperation. Comprising 26 member countries and the European Commission, the CSLF supports knowledge sharing, joint research projects, and policy development. It emphasizes the need to make CCS accessible to both industrialized and developing countries by promoting best practices, regulatory frameworks, and demonstration projects.

Another significant organization is the Global CCS Institute, a think tank and project support hub that works with governments and industries to accelerate the deployment of CCS. The Institute conducts comprehensive research, publishes annual reports on the global status of CCS, and connects stakeholders through events, knowledge hubs, and technical workshops. It also serves as an advocate for CCS within international climate discussions and works to raise awareness about its importance among policymakers and the public.

The International Energy Agency (IEA) also plays a vital role through its Clean Energy Transitions Programme, which supports CCS R&D as part of broader low-carbon energy strategies. The IEA provides policy analysis, maintains databases of global CCS projects, and convenes technical collaborations through its Greenhouse Gas R&D Programme (IEAGHG). This program produces detailed technical studies on topics such as storage site monitoring, capture technology performance, and lifecycle emissions assessment.

Key Collaborative CCS Projects around the World

Collaborative projects are vital to translating international commitments into tangible progress. In Europe, the Northern Lights project, part of Norway's Longship initiative, exemplifies the potential of multinational cooperation in CCS. The project involves the Norwegian government, Shell, Equinor, and TotalEnergies in building a shared CO_2 transport and storage infrastructure in the North Sea. It allows CO_2 captured from industrial emitters in various European countries to be shipped and permanently stored offshore, establishing a scalable model for cross-border CCS systems.

In North America, the United States and Canada have cooperated on CCS research through institutions such as the U.S. Department of Energy's National Energy Technology Laboratory and Natural Resources Canada. Projects like the Boundary Dam CCS facility in Saskatchewan and Petra Nova in Texas have benefited from shared technical expertise and joint investment in monitoring technologies and capture systems. Furthermore, both countries participate in regional knowledge networks such as the North American Carbon Capture and Storage Association, which facilitates collaboration among industries, universities, and government agencies.

In Asia, collaborative R&D is advancing through initiatives such as the Japan-Asia CCS Network, which supports capacity building, technology transfer, and project development across Southeast Asia. Japan, with its advanced capture technology and engineering capacity, provides assistance to countries like Indonesia, Malaysia, and Vietnam through feasibility studies, policy advising, and pilot project funding. These partnerships are critical for overcoming local barriers related to regulatory readiness and financing.

Technical and Regulatory Harmonization through Collaboration

One of the primary advantages of international CCS networks is the ability to harmonize technical standards and regulatory practices. For CCS to function across borders, there must be agreement on safety protocols, CO_2 quality specifications, monitoring requirements, and long-term liability frameworks. Organizations like the International Organization for Standardization (ISO) and the IEAGHG have been working on standardizing key aspects of CCS, such as site selection criteria, measurement techniques, and verification protocols.

Regulatory harmonization also supports the development of carbon markets and carbon accounting systems that recognize stored CO_2 as a verifiable emission reduction. Without such recognition, there is limited financial incentive for emitters to invest in CCS. Collaborative efforts help ensure that credits from CCS projects are compatible with voluntary and compliance markets, including the Clean Development Mechanism, Article 6 mechanisms of the Paris Agreement, and national carbon pricing schemes.

Capacity Building and Knowledge Transfer

For countries in the Global South, collaboration in CCS offers critical opportunities for capacity building. International initiatives often include training programs, technical exchanges, and support for developing local expertise in CCS science and engineering. This is particularly important for countries that lack prior experience with large-scale energy infrastructure or subsurface geology. Knowledge sharing can take the form of technical workshops, fellowships, virtual knowledge platforms, and in-country demonstration projects supported by international partners. For instance, the World Bank's CCS Trust Fund has supported feasibility studies and capacity-building activities in developing countries such as South Africa, Indonesia, and Mexico. These programs often focus on stakeholder engagement, regulatory preparedness, and geologic storage assessments. As these countries gain expertise, they become better positioned to initiate their own CCS projects or participate in regional hubs.

Financing CCS through International Partnerships

Financing remains one of the greatest challenges for CCS, especially in lower-income countries where capital constraints are more severe. Collaborative R&D and global CCS networks help unlock financing by pooling resources, sharing risk, and attracting investment from development banks, climate funds, and philanthropic organizations. The European Union, for example, supports international CCS collaboration through its Horizon Europe program, which funds joint research across member states and global partners. The role of multilateral development banks such as the Asian Development Bank (ADB), African Development Bank (AfDB), and the World Bank is also central in financing early-stage CCS research and infrastructure in developing countries. These institutions provide grants, concessional loans, and risk mitigation instruments that help de-risk investments and catalyze private sector involvement.

7.3.2 CCS Hubs and Cross-Border Projects

Role and Advantages of CCS Hubs

CCS hubs are designed to centralize the capture, transport, and storage of CO_2 from multiple industrial sources. Instead of building isolated and redundant infrastructure for each emitter, CCS hubs facilitate a networked approach that enhances scalability and cost-efficiency. The shared use of infrastructure such as pipelines, compression stations, and storage reservoirs reduces capital expenditures while enabling economies of scale. Additionally, the concentration of activities within a defined region allows for better monitoring, regulatory oversight, and public engagement. From an operational standpoint, CCS hubs can accommodate industrial clusters, including cement plants, refineries, steel mills, and chemical factories, which are typically co-located in port cities or industrial zones. This geographic proximity makes it easier to capture CO_2 emissions and transport them to nearby storage sites. The integration of capture technologies across diverse industries also allows for technological learning and innovation through shared data and collaborative R&D. In turn, this supports faster commercialization and broader deployment of CCS systems.

Examples of Operational and Planned CCS Hubs

Several CCS hubs have already been developed or are in advanced planning stages around the world. In Europe, the Northern Lights project in Norway represents a pioneering example of a cross-border CCS hub. Developed by Equinor, Shell, and TotalEnergies, Northern Lights is part of the broader Longship project and aims to transport captured CO_2 from various European countries to offshore storage sites in the North Sea. The project involves shipping CO_2 by tanker from emitters in countries such as the Netherlands, Belgium, and Germany to a receiving terminal in western Norway. From there, it is piped offshore and injected into deep saline aquifers for permanent storage. Northern Lights sets an important precedent for future cross-border CCS cooperation and is designed to expand capacity over time to accommodate additional partners.

The United Kingdom is also developing a series of CCS hubs under its "cluster sequencing" strategy. The East Coast Cluster and HyNet North West are two prominent examples, each targeting large-scale industrial emissions with the goal of becoming operational by the late 2020s. These hubs will capture CO_2 from multiple sectors and transport it via pipeline networks to offshore storage sites in the UK Continental Shelf. They demonstrate the UK's ambition to lead in CCS deployment while creating jobs and stimulating economic growth in industrial heartlands.

In North America, the United States and Canada have several existing and planned CCS hubs. The Alberta Carbon Trunk Line in Canada connects industrial emitters in central Alberta to mature oil fields where CO_2 is used for enhanced oil recovery and permanent storage. Similarly, the United States is pursuing CCS development through regional initiatives such as the CarbonSAFE program and the Houston CCS hub, which aims to capture CO_2 from the petrochemical and refining industries along the Gulf Coast.

Cross-Border Projects and Regional Cooperation

As CCS infrastructure expands, cross-border projects have become a strategic solution for countries that lack domestic storage capacity but have significant industrial emissions. These projects require extensive cooperation in terms of legal frameworks, liability agreements, permitting procedures, and infrastructure development. Europe has taken the lead in developing cross-border CCS projects, with the European Union actively supporting them through policy instruments such as the Trans-European Networks for Energy (TEN-E), the Innovation Fund, and the Connecting Europe Facility. The Porthos project in the Netherlands is a key example of regional integration. It involves capturing CO_2 from industrial emitters in the Port of Rotterdam and transporting it via pipeline to offshore storage locations in the Dutch North Sea. Although Porthos is a domestic project, it is designed to be part of a broader European CCS network that includes partnerships with Belgian and German industries, potentially extending its impact beyond national borders.

Another cross-border initiative is the cooperation between Denmark and Sweden through the Greensand project. Denmark's North Sea basin offers extensive storage potential, while Sweden has industrial emissions but limited domestic storage capacity. By enabling CO_2 to be transported across borders and injected offshore in Denmark, these countries demonstrate how shared CCS networks can overcome national limitations and foster collective climate action.

Legal, Regulatory, and Governance Challenges

While CCS hubs and cross-border projects offer considerable advantages, they also introduce complex legal and regulatory challenges. Transporting CO_2 across national boundaries requires alignment on environmental standards, safety protocols, liability regimes, and long-term monitoring responsibilities. The London Protocol, an international treaty governing marine pollution, initially prohibited the transboundary export of CO_2 for offshore storage. However, amendments adopted in recent years allow such exports under bilateral agreements, paving the way for international CCS cooperation. To facilitate these developments, countries must establish harmonized permitting processes, agree on jurisdictional responsibilities, and ensure transparent reporting systems. Liability for stored CO_2 is a particularly sensitive issue, as governments and companies must determine who is responsible in the event of leakage or failure. Addressing these concerns requires robust legal frameworks, insurance mechanisms, and institutional capacity to enforce regulations.

Global CCS Networks and International Institutions

International cooperation in CCS is supported by various organizations and networks that promote knowledge sharing, standardization, and technical collaboration. The Global CCS Institute serves as a central hub for project tracking, policy analysis, and stakeholder engagement. It maintains a global database of CCS projects, publishes annual status reports, and facilitates forums where government and industry representatives can exchange best practices.

The Carbon Sequestration Leadership Forum (CSLF), a multilateral body comprising governments, industry, and academia, also promotes collaborative research and policy alignment. It supports capacity-building efforts, especially in developing countries, and encourages the development of pilot projects and large-scale demonstrations. The International Energy Agency's Greenhouse Gas R&D Programme (IEAGHG) provides technical insights into capture, transport, and storage technologies through in-depth research reports and international workshops. These networks play a critical role in creating a global CCS community that can respond to common challenges, share innovations, and coordinate infrastructure development. By fostering trust and alignment among stakeholders, they help accelerate CCS deployment and ensure that cross-border projects are not hindered by fragmented regulatory or technical barriers.

Economic and Environmental Benefits of CCS Hubs

In addition to emissions reduction, CCS hubs offer significant economic and environmental benefits. They create high-quality jobs in engineering, construction, and maintenance while revitalizing industrial regions through clean technology investment. By enabling the decarbonization of energy-intensive sectors, hubs support the long-term competitiveness of industries in a low-carbon economy. Environmental benefits include improved air quality, reduced acidification, and the preservation of ecosystems that would otherwise be harmed by unchecked emissions.

Cross-border CCS projects also enhance energy and climate security by enabling shared responsibility for climate mitigation. They allow countries to optimize their resources—some contributing storage capacity, others providing technological expertise or financial investment. This collective approach enhances global resilience to climate risks and demonstrates how international cooperation can lead to equitable and scalable solutions.

CHAPTER

8

Future of Carbon Capture and Storage

8.1 ADVANCEMENTS IN CCS TECHNOLOGY

8.1.1 New Materials and Sorbents for CO_2 Capture

The development of new materials and sorbents for CO_2 capture is transforming the effectiveness and feasibility of Carbon Capture and Storage systems. From solid amine-based sorbents to advanced metal-organic and covalent organic frameworks, research efforts are converging on solutions that combine high capacity, selectivity, stability, and energy efficiency. While significant challenges remain in scaling production, reducing costs, and ensuring long-term performance, the progress in material science is bringing CCS closer to widespread industrial adoption. As the world intensifies its efforts to combat climate change, the innovation of new sorbents stands as a cornerstone for realizing a sustainable and low-carbon future.

Limitations of Traditional Sorbents

Traditionally, the capture of CO_2 has been carried out using liquid amine-based solvents, particularly monoethanolamine (MEA). These solvents chemically react with CO_2 in a reversible process, allowing for the regeneration and reuse of the capture medium. However, MEA-based systems have significant drawbacks. They are highly energy-intensive during the regeneration phase, where captured CO_2 must be stripped from the solvent through heating. Additionally, these solvents are corrosive, degrade over time, and require extensive maintenance and chemical replenishment, thereby increasing operational costs. The limitations of MEA and similar materials have prompted the search for alternative solid and hybrid sorbents with better performance characteristics and lower energy penalties.

Development of Solid Sorbents

Solid sorbents offer several advantages over traditional liquid systems. They can operate at lower regeneration temperatures, are less corrosive, and tend to be more stable over multiple cycles. Among solid sorbents, the most commonly studied categories include activated carbon, zeolites, metal-organic frameworks (MOFs), covalent organic frameworks (COFs), and mesoporous silicas. Each of these materials has unique structural and chemical properties that influence their CO_2

adsorption capabilities. Activated carbon is widely available and inexpensive, but it has moderate selectivity for CO_2 and limited capacity in the presence of moisture. Zeolites, on the other hand, are crystalline aluminosilicates with well-defined pore structures that allow for selective adsorption. They exhibit high surface areas and thermal stability, but their performance can be severely hampered by water vapor, which competes with CO_2 for adsorption sites. To address these issues, hydrophobic modifications and novel synthesis techniques are being explored to enhance the water resistance and capacity of zeolites.

Emergence of Metal-Organic Frameworks (MOFs)

Metal-organic frameworks have emerged as some of the most promising materials for CO_2 capture due to their highly tunable structures and exceptionally high surface areas. MOFs are composed of metal ions coordinated to organic ligands, forming porous crystalline networks with extensive internal surface areas. These materials can be chemically modified to enhance CO_2 affinity, improve selectivity over other gases, and allow for easier regeneration. Specific MOFs have demonstrated CO_2 adsorption capacities that surpass traditional materials, and many can operate under moderate conditions, reducing energy inputs. MOFs can be tailored to target specific gas mixtures and environmental conditions. For example, the incorporation of amine-functionalized ligands or the selection of metal nodes with strong interactions with CO_2 molecules can significantly enhance sorption performance. Despite their promise, however, MOFs face challenges related to stability under humid or acidic conditions, synthesis scalability, and cost. Addressing these concerns is central to bringing MOF-based technologies from the laboratory to industrial deployment.

Advancements in Covalent Organic Frameworks (COFs)

Covalent organic frameworks are another emerging class of porous crystalline materials designed with organic building blocks connected through covalent bonds. Like MOFs, COFs offer high surface areas and tunable structures, but they are composed entirely of light elements such as carbon, hydrogen, nitrogen, and oxygen, which can lead to superior thermal and chemical stability. COFs have been investigated for their ability to capture CO_2 through physisorption or chemical functionalization. One of the key advantages of COFs is their low density and potential for low-cost production using earth-abundant materials. They also allow for easier integration into membrane or mixed-matrix systems, which can be used in modular CO_2 capture units. Continued research into functionalized COFs aims to improve their selectivity, response to moisture, and recyclability, which are essential for industrial applications.

Hybrid Materials and Composite Sorbents

Researchers are also developing hybrid materials that combine the strengths of different sorbents to overcome individual limitations. For instance, integrating MOFs or COFs with polymers, graphene, or silica matrices can result in composite materials that offer both high performance and structural robustness. Such composites can be used in pressure swing adsorption or temperature swing adsorption systems with enhanced cycling capabilities. Another approach involves embedding amine groups into porous materials to create solid amine sorbents that mimic the chemical absorption process of liquid amines without the associated drawbacks. These materials show strong CO_2 affinities and good regeneration behavior, making them suitable for post-combustion applications. However, challenges remain in ensuring consistent performance under real flue gas conditions, which often include moisture, sulfur compounds, and other contaminants.

Nanotechnology and Surface Engineering

Nanotechnology has played a crucial role in the development of next-generation sorbents. Through precise control over pore size, surface functionalization, and particle morphology, researchers are designing materials with optimized interaction sites for CO_2. Surface engineering techniques allow for the selective tuning of adsorptive sites, thereby enhancing selectivity and kinetics. Nanostructured materials also exhibit faster adsorption-desorption cycles, which is critical for industrial applications requiring high throughput. Functional coatings, dopants, and heteroatom modifications are additional strategies being used to improve the CO_2 capture properties of materials. These techniques are enabling the development of "smart" sorbents that respond dynamically to temperature, pressure, or gas composition changes, allowing for more efficient system integration.

Scalability, Cost, and Environmental Considerations

While many new materials show excellent laboratory performance, scalability and cost remain major hurdles for real-world deployment. The synthesis of MOFs and COFs, for instance, often involves complex procedures and expensive precursors. Therefore, efforts are underway to develop more sustainable synthesis routes, including green chemistry approaches, solvent-free reactions, and the use of biomass-derived components. Life cycle assessments are being conducted to evaluate the environmental impacts of sorbent production, usage, and disposal, ensuring that CCS technologies remain environmentally beneficial. Moreover, durability under operational conditions is a key criterion for commercial adoption. Sorbents must withstand thousands of cycles of adsorption and desorption without significant degradation. Long-term testing under realistic conditions, including exposure to impurities and variable temperatures, is necessary to validate the practical potential of new materials.

Integration with Industrial Processes and Modular Design

The integration of advanced sorbents into existing industrial processes is a critical aspect of commercialization. New materials must be compatible with current infrastructure or enable the design of modular systems that can be easily installed and maintained. Modular units based on solid sorbents are being developed for point-source applications, such as on-site capture at smaller facilities or distributed energy systems. These units benefit from the compact nature and high performance of solid sorbents, especially those derived from nanostructured or functionalized materials. Additionally, advances in computational modeling and artificial intelligence are aiding the design and screening of new sorbents. Machine learning algorithms can predict material behavior, optimize compositions, and accelerate the discovery process. This computational-materials approach reduces experimental time and costs while expanding the range of potential candidates for CO_2 capture.

8.1.2 Breakthroughs in Transport and Storage Solutions

Breakthroughs in transport and storage solutions are pivotal to the success of Carbon Capture and Storage as a climate mitigation technology. Advances in pipeline and shipping technologies, real-time monitoring, and digital infrastructure are improving the efficiency and safety of CO_2 transport. Meanwhile, innovations in geological modeling, offshore storage, and post-injection monitoring are enhancing the reliability of long-term storage. These developments, supported by robust policy frameworks and international collaboration, are laying the foundation for a global CCS network that can significantly reduce greenhouse gas emissions and help achieve climate goals. As these breakthroughs continue, CCS is increasingly positioned as a scalable, secure, and essential component of the low-carbon transition.

Innovations in CO_2 Transport Infrastructure

The transport phase of CCS traditionally involves moving compressed carbon dioxide from the capture site to a storage location through pipelines or, in some cases, ships and trucks. Pipelines have been the most commonly used method due to their efficiency in handling large volumes over long distances. Breakthroughs in pipeline technology have focused on optimizing materials for corrosion resistance, developing advanced monitoring systems, and expanding networks to serve multiple capture sources. Modern pipelines are now being constructed with advanced alloys and composite materials that reduce the risk of leaks and degradation caused by the acidic nature of CO_2, especially when water is present. Another major development is the concept of shared or "backbone" CO_2 pipeline networks. These infrastructures are designed to serve multiple emitters and connect them with large, centralized storage facilities. By reducing duplication and increasing economies of scale, shared pipelines make CCS more economically viable for smaller emitters and industrial clusters. Countries like Norway, the Netherlands, and the United Kingdom are leading in the implementation of such shared infrastructure, supported by public-private partnerships and regional planning.

In addition to pipelines, there has been increased attention to shipping CO_2, particularly for projects where pipeline construction is not feasible due to geographical constraints. Liquefied CO_2 ships, similar to LNG carriers, are being developed and tested for long-distance maritime transport. These vessels require advanced insulation and compression systems to keep CO_2 in liquid form, and innovations in ship design are making this mode of transport more efficient and safer. Shipping provides flexibility and allows CCS projects to access distant offshore storage sites, facilitating international cooperation in carbon storage.

Digitalization and Real-Time Monitoring in Transport

One of the transformative advances in CO_2 transport is the integration of digital technologies for monitoring and maintenance. Real-time sensors embedded along pipelines can detect pressure changes, flow rates, and potential leaks, enhancing safety and enabling rapid response to anomalies. These sensors are connected to centralized control systems using the Internet of Things (IoT), allowing operators to monitor pipeline performance remotely and in real time. Predictive maintenance technologies, powered by artificial intelligence and machine learning, are also being applied to anticipate wear and optimize inspection schedules, reducing downtime and operational costs. Moreover, Geographic Information Systems (GIS) are used for planning and routing pipeline infrastructure, considering environmental impact, terrain, land use, and seismic risk. These digital tools enhance the planning phase and support regulatory compliance while ensuring that pipelines avoid sensitive areas. Such innovations are crucial for gaining public and political acceptance, which is often a hurdle in infrastructure projects.

Advances in CO_2 Compression and Conditioning

Before CO_2 can be transported, it must be compressed and conditioned to a dense phase, typically as a supercritical fluid. Recent breakthroughs in compression technology have focused on energy efficiency and modular design. Multi-stage compressors with intercooling and improved sealing systems reduce energy requirements and emissions from the compression process itself. Modular compression units allow for scalable deployment, making it easier to tailor systems for different industrial scales. Another important development is the improvement of CO_2 dehydration systems. The presence of water in CO_2 streams can lead to corrosion and hydrate formation in pipelines, which poses safety and reliability risks. Modern dehydration units use advanced membranes and desiccants to ensure high levels of purity before transport. These systems are being refined to operate efficiently at varying pressures and CO_2 concentrations, increasing their adaptability to diverse industrial processes.

Breakthroughs in Geological Storage Techniques

The long-term storage of CO_2 relies primarily on geological formations such as depleted oil and gas reservoirs, deep saline aquifers, and unmineable coal seams. Over the past two decades, advancements in subsurface imaging, modeling, and monitoring have significantly improved the safety and reliability of underground storage. Seismic imaging technologies, including 4D seismic surveys, allow operators to track the movement of injected CO_2 over time, ensuring it remains contained within the designated reservoir. One major breakthrough is the development of sophisticated reservoir simulation software that integrates geological, chemical, and mechanical data to predict CO_2 behavior post-injection. These models help identify optimal injection sites, estimate storage capacity, and assess risks such as caprock failure or induced seismicity. Enhanced modeling tools also support dynamic monitoring plans, allowing adjustments to injection strategies based on real-time subsurface conditions.

The deployment of pressure management systems in storage formations is another significant advancement. By using brine extraction or injecting fluids in neighboring wells, operators can manage reservoir pressure more effectively and minimize the risk of fault activation or caprock breach. These techniques improve the integrity of storage sites and expand the range of viable geological formations.

Offshore Storage and Integrated Projects

Offshore storage is gaining momentum as a preferred option due to the vast capacity available beneath continental shelves and the reduced likelihood of public opposition compared to onshore sites. Projects like Norway's Sleipner and the Northern Lights initiative demonstrate the viability of offshore saline aquifer storage. Innovations in offshore well design, marine monitoring systems, and subsea infrastructure have made these projects increasingly attractive and replicable. Integrated CCS projects that combine capture, transport, and storage across regions are being developed to take advantage of offshore resources. These initiatives often include multiple emitters and cross-border cooperation, aligning with climate goals and promoting shared investment. For instance, the Northern Lights project involves cooperation between Norway, the European Union, and private sector actors to develop a comprehensive CCS value chain, from industrial capture facilities to offshore injection wells.

Storage Site Monitoring and Post-Injection Management

Ensuring the long-term containment of CO_2 is essential for the environmental and regulatory acceptance of CCS. Breakthroughs in monitoring technologies have greatly enhanced the ability to track CO_2 plumes and detect any potential leakage. Fiber optic sensors, microseismic monitoring, and satellite-based remote sensing provide high-resolution data on subsurface changes and surface deformation. These systems are often coupled with machine learning algorithms to analyze data and detect anomalies that could indicate a leak. In addition to subsurface monitoring, surface soil gas surveys and groundwater sampling are used to detect early signs of CO_2 migration.

The integration of multiple monitoring methods, referred to as "monitoring, verification, and accounting" (MVA), is becoming the standard for CCS projects worldwide. This comprehensive approach not only ensures safety but also supports regulatory compliance and carbon credit verification. Post-injection site care is also evolving, with strategies to transition monitoring responsibilities from operators to public agencies over time. These frameworks include contingency plans, liability coverage, and long-term data archiving to maintain confidence in the integrity of storage sites for decades after closure.

Policy and Regulatory Support for Transport and Storage

The technological breakthroughs in CO_2 transport and storage are closely linked to evolving policy and regulatory frameworks. Governments are introducing streamlined permitting processes, financial incentives, and cross-border agreements to facilitate the deployment of CCS infrastructure. In the European Union, the TEN-E regulation supports cross-border CO_2 infrastructure, while the United States provides tax credits under Section 45Q for stored and utilized CO_2. These policies encourage investment and collaboration, which are critical for scaling up CCS networks. Legal clarity around long-term liability, pore space rights, and environmental standards is also improving, reducing uncertainty for project developers. Public engagement strategies and transparent communication are increasingly integrated into project planning to address concerns and build trust among stakeholders.

8.1.3 Integration of CCS with Renewable Energy Systems

The primary rationale for integrating CCS with renewable energy systems lies in the complementary strengths of each technology. Renewable energy sources such as solar and wind are intermittent and variable, depending on weather conditions and time of day. As a result, they require backup power or energy storage solutions to ensure grid stability. CCS, especially when applied to bioenergy (BECCS) or hydrogen production, can provide dispatchable power that fills in the gaps during periods of low renewable generation. By capturing and storing CO_2 from these processes, the overall carbon intensity of the energy system is reduced, creating a synergistic effect that enhances climate mitigation efforts.

One of the most promising areas for integration is the combination of CCS with bioenergy, known as BECCS. In this system, biomass such as agricultural residues or forestry byproducts is used to produce electricity or fuels, and the CO_2 released during combustion or processing is captured and stored underground. Because the biomass absorbs CO_2 from the atmosphere as it grows, capturing the carbon emissions during its use creates a net-negative emissions process. BECCS is widely recognized in climate models as a key technology for achieving global temperature targets and compensating for residual emissions in hard-to-abate sectors. When paired with renewable electricity inputs for biomass processing, the climate benefits are further magnified.

Another significant opportunity lies in the integration of CCS with hydrogen production. Hydrogen is gaining global attention as a clean energy carrier that can decarbonize transportation, industry, and heating. While green hydrogen, produced using electrolysis powered by renewable energy, is the most sustainable form, it remains expensive and constrained by limited electrolyzer capacity. Blue hydrogen, derived from natural gas reforming with CCS, offers a more immediately scalable solution. By capturing CO_2 from the reforming process, blue hydrogen provides a low-carbon alternative that can be deployed rapidly and at scale. This form of integration supports the expansion of the hydrogen economy while building the infrastructure necessary for future green hydrogen systems.

Energy storage is another area where CCS can complement renewable energy systems. During periods of excess renewable generation, surplus electricity can be used to power direct air capture (DAC) systems that remove CO_2 from the atmosphere. The captured carbon can then be stored permanently or utilized for industrial purposes. This coupling of renewable power with DAC and storage creates a form of flexible negative emissions capacity, enhancing grid resilience and climate outcomes. The concept of "power-to-carbon removal" is gaining attention as a potential grid service, especially in regions with high shares of renewables and significant seasonal variability in energy supply.

Geographical considerations also support the integration of CCS with renewables. Many regions with strong renewable resources, such as the North Sea, Australia, and parts of North America, also have access to suitable geological formations for CO_2 storage. This co-location allows for efficient development of integrated energy systems that combine renewable generation, carbon capture, and underground storage. Infrastructure synergies, including shared pipelines, storage hubs, and port facilities, further reduce costs and facilitate coordinated planning across sectors.

Policy frameworks are beginning to recognize the value of integration. For example, the European Union's Green Deal and Fit for 55 package emphasize the role of BECCS and blue hydrogen in decarbonization strategies. In the United States, tax credits under Section 45Q incentivize both CCS and renewable energy deployment, encouraging hybrid projects that combine both elements. The development of carbon markets and renewable energy certificates can also support integrated projects by monetizing both emissions reductions and clean energy generation. Regulatory harmonization across sectors and jurisdictions will be critical to unlocking the full potential of these synergies.

From a technological standpoint, integration presents both challenges and opportunities. System design must account for the different operational characteristics of renewable sources and CCS facilities. While solar and wind operate on variable output, CCS systems often require steady-state conditions for optimal performance. Innovative solutions such as hybrid plants with flexible operation modes, energy storage integration, and advanced control systems are being developed to bridge this gap. Research into modular and scalable CCS technologies is also enabling more flexible deployment that can adapt to the needs of renewable-dominated grids.

Public perception and community engagement are essential to the successful integration of CCS and renewables. While renewable energy generally enjoys strong public support, CCS projects can face opposition due to concerns about safety, environmental risks, and proximity to communities. Transparent communication, community benefits, and stakeholder involvement are key to building trust and social acceptance. Integrated projects that demonstrate clear climate and economic benefits are more likely to gain public backing and political support.

The economic dimension of integration is multifaceted. While standalone CCS projects can be capital-intensive, coupling them with renewable energy can improve economic viability through shared infrastructure and diversified revenue streams. For example, a biomass power plant with CCS can sell electricity, carbon removal credits, and possibly biofuels, creating multiple income sources. Similarly, hydrogen production facilities with CCS can supply industrial gas, participate in carbon markets, and contribute to energy security. Cost-sharing across these functions improves the return on investment and makes integrated systems more attractive to investors.

Research and development are playing a vital role in advancing integration. Pilot projects and demonstration plants are testing new configurations, optimizing process efficiency, and validating performance under real-world conditions. International collaboration through organizations like Mission Innovation and the Clean Energy Ministerial is accelerating knowledge sharing and technology diffusion. Academic and industrial partnerships are also fostering innovation in carbon utilization, renewable integration, and digital monitoring systems. As these efforts mature, they will help bring integrated CCS-renewable systems closer to commercial scale. Looking ahead, the role of integrated systems will become increasingly important as countries strive to meet climate targets. Scenarios developed by the Intergovernmental Panel on Climate Change (IPCC) and the International Energy Agency (IEA) highlight the need for both CCS and renewables in a balanced energy transition. Integrated systems provide the flexibility, scalability, and carbon reduction potential required to support deep decarbonization. They also offer pathways for sectors that are difficult to electrify, such as heavy industry, aviation, and long-distance transport.

8.2 THE ROLE OF CCS IN ACHIEVING NET-ZERO EMISSIONS BY 2050

8.2.1 Global Scenarios for CCS Deployment

The global commitment to achieving net-zero greenhouse gas emissions by 2050 is a response to the alarming progression of climate change. Within this effort, Carbon Capture and Storage (CCS) technologies have emerged as one of the most vital solutions. CCS encompasses the processes of capturing carbon dioxide from high-emission sources such as power plants and industrial operations, transporting the captured carbon to designated sites, and injecting it into deep underground geological formations for long-term storage. This technology enables the reduction of atmospheric carbon dioxide, thereby playing a central role in mitigation strategies outlined in many net-zero roadmaps. CCS is particularly crucial for sectors where full decarbonization is technologically or economically infeasible by mid-century, such as cement, steel, and petrochemicals.

Achieving net-zero emissions by 2050 is an ambitious goal that requires the deployment of every available mitigation tool. In this context, CCS stands as a crucial component of the global climate strategy. It offers a solution for sectors with limited alternatives, enables negative emissions, and facilitates the decarbonization of power and industry. While challenges related to cost, infrastructure, public acceptance, and policy remain, they are surmountable with political will, international cooperation, and sustained innovation. The global deployment of CCS must be accelerated and integrated into national climate action plans if the world is to meet the 2050 net-zero target. The years ahead will be critical, and the decisions made now will determine the role that CCS ultimately plays in shaping a sustainable, low-carbon future.

CCS in Global Emission Reduction Scenarios

Global models and climate projections highlight the importance of CCS in achieving climate goals. The Intergovernmental Panel on Climate Change (IPCC) has emphasized that most 1.5°C and 2°C mitigation pathways rely significantly on CCS. Scenarios developed by the International Energy Agency (IEA) similarly forecast that CCS could contribute to approximately 15 percent of the cumulative emissions reductions required by 2050. This role becomes even more critical as emissions from aviation, heavy industry, and agriculture are unlikely to be fully eliminated, making negative emissions technologies such as Bioenergy with Carbon Capture and Storage (BECCS) and Direct Air Capture (DAC) essential. CCS, therefore, is not just a transition solution but a long-term necessity in the broader climate mitigation architecture.

Advancements and Deployment in Developed Economies

In high-income countries, CCS has moved from the theoretical stage to practical implementation. Projects such as Norway's Sleipner, Canada's Boundary Dam, and the United States' Petra Nova have demonstrated the technical viability of capturing and storing carbon dioxide on a large scale. These projects have benefited from supportive policies, financial incentives, and established regulatory frameworks. For example, the United States has implemented the 45Q tax credit to stimulate private investment in CCS by offering financial returns for each ton of carbon dioxide captured and stored. In the European Union, CCS is being integrated into broader strategies for achieving carbon neutrality, supported by initiatives like the European Green Deal and funding under the Innovation Fund.

The Role of CCS in Emerging and Developing Economies

Emerging economies, particularly China and India, also recognize the potential of CCS, although deployment is still limited compared to developed nations. China has begun to incorporate CCS in its national development plans and has launched pilot projects in the power and industrial sectors. In the Middle East, countries like Saudi Arabia and the United Arab Emirates are investing in CCS as part of their broader goal of developing a circular carbon economy. However, in many developing countries, CCS adoption is constrained by high capital costs, lack of technical expertise, and competing socioeconomic priorities. For these countries, international cooperation, technology transfer, and climate finance mechanisms will be essential to unlock the full potential of CCS.

Infrastructure and Storage Considerations

The expansion of CCS technologies globally requires a corresponding development of CO_2 transport and storage infrastructure. This involves the construction of pipelines, conversion of shipping terminals, and identification of secure geological storage sites such as deep saline aquifers, basalt formations, and depleted oil and gas fields. The success of CCS depends on this logistical backbone, which must be planned with foresight and coordination. Europe's Northern Lights project, which provides open-access CO_2 transport and storage services to multiple emitters, represents a pioneering approach to shared infrastructure. This model can be replicated in other regions to enable cost-efficient and widespread CCS deployment.

Technological Innovations Driving CCS Efficiency

Advancements in CCS technology are reducing the cost and increasing the efficiency of carbon capture. New solvents and sorbents are making chemical absorption more energy-efficient, while innovations in membrane technologies promise to revolutionize separation processes. Solid oxide electrolysis and other next-generation capture techniques are also under development. Additionally, the rise of Direct Air Capture, though currently expensive and energy-intensive, presents opportunities for removing carbon dioxide directly from the atmosphere. With sufficient investment in research and development, the economic and technical feasibility of both point-source capture and negative emissions technologies will continue to improve, accelerating their integration into climate strategies.

Economic Viability and Investment Challenges

The large-scale deployment of CCS hinges on economic feasibility. Despite technological readiness, most CCS projects remain capital-intensive and face long payback periods. Government subsidies, carbon pricing, and financial instruments such as contracts-for-difference can help de-risk investment and make CCS more attractive to the private sector. Voluntary carbon markets and emerging compliance markets are also beginning to recognize carbon removals via CCS, which could further incentivize development. Nevertheless, a significant gap remains between the level of investment needed and current funding commitments. Closing this gap will require coordinated action from governments, multilateral institutions, and the private sector.

Societal Acceptance and Regulatory Safeguards

Widespread adoption of CCS also depends on public perception and social license to operate. Concerns about the long-term safety of underground storage, potential leakage, and environmental justice must be addressed through transparent communication and stringent regulatory oversight. Public engagement strategies, environmental impact assessments, and monitoring protocols are essential for building trust. International standards, such as those developed by the International Organization for Standardization (ISO) and the Carbon Sequestration Leadership Forum (CSLF), help ensure the safe and responsible management of storage sites across jurisdictions.

Enabling a Hydrogen and Low-Carbon Industrial Future

CCS is also critical in enabling low-carbon hydrogen production through a process known as blue hydrogen, where natural gas is reformed into hydrogen and carbon dioxide is captured and stored. This technology serves as a bridge to green hydrogen by allowing countries to scale hydrogen supply quickly while minimizing emissions. In addition, CCS can decarbonize other hard-to-electrify industrial sectors, helping maintain competitiveness and protect jobs while contributing to climate goals. By integrating CCS into industrial hubs and carbon-intensive clusters, countries can create synergies that enhance both economic resilience and environmental sustainability.

8.2.2 CCS in Hard-to-Decarbonize Sectors: Cement, Steel, and Aviation

As the world moves toward achieving net-zero emissions by 2050, certain industrial and transport sectors remain particularly resistant to decarbonization due to their inherent processes, energy intensity, or technological limitations. Among these hard-to-decarbonize sectors, cement, steel, and aviation stand out for their significant contributions to global carbon dioxide emissions and the difficulty of replacing their fossil fuel dependence with renewable alternatives. Carbon Capture and Storage (CCS) offers a vital technological pathway to reduce emissions from these sectors by capturing CO_2 either at the point of emission or, in some cases, directly from the atmosphere.

Cement, steel, and aviation are among the most difficult sectors to decarbonize due to their technical complexity, energy intensity, and reliance on specific chemical or physical processes. In each of these sectors, CCS provides a strategic solution that addresses the limitations of other decarbonization approaches. Whether by capturing process emissions from cement kilns, enabling low-carbon steelmaking, or producing synthetic aviation fuels, CCS is a versatile and necessary component of the global climate mitigation toolkit.

While challenges remain in terms of cost, infrastructure, and public acceptance, the successful application of CCS in these sectors could significantly reduce global greenhouse gas emissions and pave the way for a more sustainable industrial and transport system. Accelerated deployment, driven by innovation, supportive policy frameworks, and international collaboration, will be essential if CCS is to fulfill its potential in transforming some of the most carbon-intensive areas of the global economy. Achieving net-zero by 2050 will require solutions tailored to the unique characteristics of each sector, and in the case of cement, steel, and aviation, CCS stands out as both a viable and vital option.

Cement Industry and Process Emissions

The cement industry is responsible for nearly seven percent of global carbon dioxide emissions. A large portion of these emissions arises not from energy use but from the chemical process of calcination, where limestone (calcium carbonate) is heated to produce lime (calcium oxide), releasing CO_2 as a byproduct. These are known as process emissions and cannot be eliminated by switching to renewable energy sources. As a result, the decarbonization of cement production demands technologies that address both energy-related and process-related emissions. CCS is one of the only available technologies that can capture these unavoidable emissions. CCS in the cement sector involves installing capture units at the kiln exhaust, where the majority of emissions occur. Technologies such as oxy-fuel combustion and post-combustion capture using chemical solvents are being tested and deployed in pilot projects worldwide. For instance, the Norcem project in Norway is integrating CCS into a commercial cement plant with the goal of storing captured carbon in offshore geological formations. Similar initiatives are underway in Europe, Canada, and parts of Asia. The implementation of CCS in cement production presents challenges such as high retrofit costs, the need for continuous CO_2 capture due to uninterrupted production cycles, and access to nearby storage sites or transport infrastructure. However, with appropriate policy incentives and technological advancements, CCS can substantially reduce emissions in the cement industry without altering its fundamental production processes.

Steel Production and the Limits of Electrification

The steel industry is another major emitter, accounting for around eight percent of global CO_2 emissions. Most steel is currently produced using blast furnaces, which rely on coke (derived from coal) as both a fuel and a reducing agent to convert iron ore into molten iron. This process generates significant emissions that are difficult to eliminate through electrification alone. Although alternative methods such as hydrogen-based Direct Reduced Iron (DRI) are being developed, their widespread adoption faces barriers related to cost, hydrogen supply, and infrastructure readiness.

In this context, CCS offers a practical solution to reduce emissions from existing blast furnace operations. By capturing CO_2 from flue gases, CCS can be retrofitted to current facilities, allowing them to continue operating while significantly lowering their carbon footprint. Projects like the STEPWISE initiative in Europe and the Emirates Steel Industries CCS project in the United Arab Emirates demonstrate how capture technology can be adapted to the specific conditions of steel plants. Post-combustion capture and gas reforming with carbon removal are among the techniques showing promise. As with cement, the successful application of CCS in steel requires an integrated approach that combines technological readiness with transport and storage infrastructure and favorable policy environments.

Moreover, CCS can complement the transition to low-carbon steelmaking by bridging the gap until green hydrogen becomes economically and technically viable on a large scale. Hybrid approaches that combine CCS with hydrogen or biomass are being explored to provide transitional pathways that reduce emissions while maintaining industrial output. In emerging economies, where the demand for steel is growing rapidly and the installed base of traditional blast furnaces is large, CCS may be the most feasible near-term option for decarbonization.

Aviation and the Challenge of Fuel Substitution

Among hard-to-decarbonize sectors, aviation presents perhaps the most complex challenge due to the high energy density required for long-haul flights and the current lack of scalable alternatives to liquid jet fuel. While improvements in aircraft efficiency, air traffic management, and the use of Sustainable Aviation Fuels (SAFs) offer incremental emission reductions, they are unlikely to achieve deep decarbonization on their own. SAFs, although renewable, still emit CO_2 when burned and are currently limited in availability and cost-effectiveness.

CCS can play an indirect yet crucial role in decarbonizing aviation by enabling the production of synthetic fuels, also known as e-fuels or electrofuels. These fuels are created by combining green hydrogen with carbon dioxide captured either from industrial sources or directly from the atmosphere through Direct Air Capture (DAC). The resulting liquid fuels can be used in existing aircraft engines, making them a drop-in replacement for conventional jet fuel. The use of DAC in this context allows for the carbon released during combustion to be balanced by carbon removed during fuel production, achieving net-zero emissions over the fuel life cycle.

In addition to synthetic fuel production, CCS can be used at airports and maintenance facilities to capture emissions from ground operations, though this represents a smaller share of total aviation emissions. The primary impact of CCS in aviation, therefore, lies in its integration into broader carbon management systems that support sustainable fuel development. Investments in infrastructure, carbon pricing, and subsidies for SAFs and synthetic fuels will be necessary to scale these solutions and bring them to market.

Policy Support and Economic Considerations

The deployment of CCS in cement, steel, and aviation requires significant capital investment, long-term planning, and robust policy support. Carbon pricing mechanisms, such as carbon taxes or emissions trading systems, can help internalize the environmental cost of emissions and make CCS economically competitive. Public funding for research, development, and demonstration projects is also essential, especially in the early stages of technology deployment. Regulatory frameworks must be developed to ensure the safe transport and storage of carbon dioxide and to provide clarity on liability and monitorng responsibilities. Governments can also play a role in creating demand for low-carbon materials and fuels through public procurement, building codes, and standards. For example, requiring the use of low-emission cement and steel in infrastructure projects can stimulate market demand and encourage industry investment in CCS. Similarly, aviation regulators can mandate increasing shares of SAFs or synthetic fuels, creating incentives for investment in CCS-enabled fuel production. International cooperation is vital, especially in sectors like aviation, which operate across borders and require harmonized standards and regulations.

8.3 PUBLIC AND POLITICAL WILL FOR SCALING CCS

8.3.1 Mobilizing Policy Support for CCS Expansion

Carbon Capture and Storage is a cornerstone technology for achieving deep decarbonization in sectors where emissions are otherwise unavoidable. Its successful deployment depends not only on technical feasibility but also, and more crucially, on enabling policy environments. Mobilizing policy support requires a comprehensive and multi-layered approach that encompasses regulatory clarity, financial incentives, infrastructure development, market integration, innovation support, and public engagement. Governments, in collaboration with industry, civil society, and international partners, must act decisively to mainstream CCS into national and global climate strategies. The scale and urgency of the climate challenge demand bold and coordinated policy action. Without sustained political will and targeted policy frameworks, CCS will remain underutilized and its potential unrealized. However, with the right support mechanisms, CCS can evolve from a niche solution into a globally scaled climate technology, capable of making a meaningful contribution to a net-zero future.

Creating Clear and Stable Regulatory Frameworks

A fundamental requirement for CCS expansion is the establishment of clear, stable, and coherent regulatory frameworks. These frameworks must govern all aspects of the CCS chain, including site selection, carbon capture technologies, transportation networks, storage site monitoring, and long-term liability. Inconsistent or underdeveloped regulations create uncertainty for investors and developers, hindering progress. Countries that have successfully launched CCS projects, such as Norway, the United Kingdom, and the United States, have benefited from robust legal systems that provide guidance on ownership of pore space, injection permits, and environmental safety standards. For emerging and developing economies seeking to enter the CCS space, policy clarity is especially important in order to build trust among stakeholders and attract international investment. Transparent permitting procedures, risk assessment standards, and protocols for storage site closure and post-injection monitoring are essential components of a credible regulatory architecture. These frameworks must also be adaptive to incorporate technological advances and ensure that legal structures evolve in tandem with scientific understanding.

Providing Financial Incentives and Risk Mitigation Tools

The high upfront capital costs associated with CCS projects present a significant barrier to deployment. To overcome this, governments must introduce financial incentives that can de-risk investment and make projects more economically viable. Tax credits, subsidies, and grants are among the most effective tools in this regard. The United States' Section 45Q tax credit, for example, has proven instrumental in promoting CCS projects by providing financial returns for each ton of CO_2 captured and stored or utilized. Similarly, the European Union's Innovation Fund and Connecting Europe Facility support large-scale CCS initiatives across member states. In addition to direct financial support, risk mitigation instruments such as loan guarantees, insurance schemes, and public-private partnerships can encourage private sector participation. These tools help absorb the financial risk associated with long payback periods, technology performance, and uncertain carbon pricing. Multilateral development banks and international climate finance mechanisms also have a role to play in extending concessional finance and technical assistance, especially for projects in low- and middle-income countries. By blending public and private capital through innovative financing models, policy can significantly enhance the bankability of CCS ventures.

Integrating CCS into Nationally Determined Contributions and Climate Strategies

National climate policies and commitments under the Paris Agreement provide an important entry point for CCS integration. Many countries have yet to explicitly include CCS in their Nationally Determined Contributions (NDCs) or long-term low-emissions development strategies. This omission reflects both technical uncertainties and a lack of political prioritization. To correct this, policymakers must recognize CCS as a legitimate and necessary component of national decarbonization pathways, particularly for sectors with limited alternatives such as cement, steel, and chemical manufacturing. Incorporating CCS into national climate planning not only signals political commitment but also enables governments to align regulatory, fiscal, and infrastructural policies around a coherent vision for low-carbon development. This alignment facilitates coordination among ministries, accelerates permitting processes, and ensures that CCS is integrated into broader energy, industrial, and environmental policies. It also strengthens a country's position in international negotiations and opens access to climate finance channels that support emissions reduction and technology transfer.

Enabling CO_2 Transport and Storage Infrastructure Development

One of the defining features of CCS is its dependence on interconnected infrastructure for CO_2 transport and storage. Without pipelines, shipping terminals, and accessible geological storage sites, CCS deployment remains geographically constrained and economically impractical. Public policy has a central role in facilitating the planning, financing, and permitting of this infrastructure. Just as electricity grids and water systems were developed with government backing, CO_2 networks require similar foundational support. Policymakers must adopt a systems-level approach to infrastructure development, identifying strategic storage hubs, enabling shared pipeline networks, and encouraging clustering of emitters to reduce costs through economies of scale. The concept of CCS hubs and clusters, where multiple industries share capture and transport infrastructure leading to centralized storage sites, is gaining traction in regions such as the North Sea, Gulf Coast, and Alberta. National and regional governments can catalyze such developments through infrastructure masterplans, public investments, and streamlined regulations. Enabling frameworks for cross-border transport and storage are also crucial, particularly in regions where storage capacity is unevenly distributed.

Encouraging Innovation and Technological Advancement

Continued research and innovation are vital for reducing the costs and improving the efficiency of CCS technologies. Policy support in the form of research grants, demonstration funding, and technology incubators can accelerate the development of next-generation capture techniques, such as advanced solvents, membranes, and direct air capture. Support for pilot projects allows companies to test technologies at scale and de-risk them before commercial deployment. Moreover, policy can foster collaboration between academia, industry, and government research institutions to create innovation ecosystems around CCS. Intellectual property protections, open data platforms, and knowledge-sharing networks can further enhance technological progress and ensure that breakthroughs benefit the broader community. Policies that support workforce training and education will also help develop the skilled labor force necessary for building and operating CCS facilities.

Aligning Carbon Pricing and Market Mechanisms

Carbon pricing mechanisms such as carbon taxes and emissions trading systems are powerful policy tools that can make CCS economically attractive. By assigning a cost to carbon emissions, these mechanisms create a financial incentive for emitters to adopt carbon-reducing technologies. When the cost of emitting CO_2 exceeds the cost of capture and storage, CCS becomes a rational business decision. However, for this dynamic to work effectively, carbon prices must be predictable, robust, and sufficiently high to reflect the social cost of carbon. Governments can further align CCS with market-based instruments by allowing captured CO_2 to generate tradeable emissions credits. In voluntary carbon markets, CCS-based removals such as bioenergy with carbon capture and storage (BECCS) and direct air capture (DAC) are beginning to gain recognition. Establishing clear methodologies, robust monitoring protocols, and third-party verification systems can build trust in these credits and attract investment. Inclusion of CCS in compliance markets, where emitters are mandated to reduce emissions or purchase allowances, can also create stable demand for the technology.

Building Public Trust and Social License to Operate

Public acceptance is another essential component of CCS expansion, especially when projects involve onshore storage or pass through populated areas. Misunderstandings about safety, environmental impact, and long-term risks can generate opposition that delays or halts projects. Policymakers must take a proactive role in educating the public, engaging local communities, and promoting transparency in decision-making processes. Policies should mandate community consultation, environmental impact assessments, and ongoing monitoring of storage sites. Sharing the benefits of CCS projects, such as job creation and regional development, can also help build public support. Additionally, the inclusion of civil society in planning and oversight can enhance accountability and ensure that environmental justice considerations are addressed.

8.3.2 Addressing Political and Social Challenges to Widespread Deployment

The successful deployment of Carbon Capture and Storage technologies depends not only on technical readiness and economic feasibility but also on overcoming the complex political and social challenges that shape public discourse and policy action. These challenges—ranging from political inconsistency and regulatory fragmentation to public opposition and equity concerns—require sustained engagement, transparent governance, and inclusive planning. By institutionalizing long-term policy support, fostering intergovernmental coordination, and proactively engaging with communities and civil society, governments can build the trust and legitimacy needed to scale CCS responsibly. Moreover, embedding equity, justice, and environmental integrity into every stage of CCS development will ensure that the technology supports a truly sustainable and socially acceptable path to net-zero. As climate deadlines approach, addressing these political and social barriers is not optional—it is essential to making CCS a viable, trusted, and widely accepted solution.

Problem of Political Will and Long-Term Vision

One of the most significant political challenges to widespread CCS deployment is the inconsistent level of political will across countries and within different levels of governance. Although CCS has been included in many international assessments and national strategies, it often lacks the strong, sustained political commitment needed to secure long-term funding and regulatory support. Political leaders may prioritize shorter-term, more visible climate actions, such as renewable energy expansion or electric vehicle subsidies, over less tangible investments in CCS infrastructure. The problem is exacerbated by the time lag between investment and visible environmental benefits. Unlike solar panels or wind turbines, which can demonstrate emission reductions quickly and publicly, CCS operations primarily occur underground and out of sight, making it harder to gain political capital. The absence of immediate, measurable outcomes can deter politicians from championing CCS, especially in democratic contexts where electoral cycles shape decision-making. Furthermore, without a long-term, cross-party consensus on climate strategies, CCS risks becoming a politically contentious issue. Changes in government may lead to abrupt shifts in policy, jeopardizing ongoing projects and discouraging private investment. To overcome this, countries need to institutionalize support for CCS within broader legislative frameworks, such as climate laws, national energy plans, or emissions reduction mandates. Long-term political vision, informed by scientific evidence and insulated from partisan shifts, is essential for the steady expansion of CCS.

Intergovernmental Coordination and Regulatory Fragmentation

The deployment of CCS technologies involves coordination across multiple sectors and jurisdictions, including energy, environment, industry, and land use planning. This makes the regulatory landscape inherently complex and fragmented, particularly in federal systems where authority is divided between national and sub-national governments. Misalignment between agencies can result in bureaucratic delays, conflicting standards, and regulatory gaps that stall CCS development.

For instance, while a national government might promote CCS through subsidies or tax credits, state or local authorities may resist the construction of pipelines or storage facilities due to environmental or political concerns. Similarly, divergent regulations across borders can impede the creation of regional CCS networks that rely on shared infrastructure. These issues highlight the need for integrated governance models that streamline regulatory processes, harmonize standards, and facilitate information-sharing between institutions.

Policymakers must also consider cross-border collaboration for countries with limited geological storage capacity but high industrial emissions. Establishing international agreements and legal protocols for transboundary CO_2 transport and storage will be vital in ensuring that regional CCS networks are legally and operationally feasible. Coordinated policies and shared regulatory frameworks will not only improve efficiency but also build trust among stakeholders.

Public Perception and Social Acceptance

Social acceptance is another major barrier to the large-scale deployment of CCS. Public skepticism often stems from a lack of awareness, misunderstanding of the technology, or concerns about safety, environmental impacts, and the motives of project developers. In some cases, CCS is viewed as a strategy to prolong the use of fossil fuels, leading to opposition from environmental groups who argue that it undermines the transition to renewable energy.

Public resistance can be especially strong in communities near proposed storage sites or pipeline routes. Concerns about groundwater contamination, induced seismicity, or leaks can trigger local opposition, delay permitting, and result in costly legal battles. These fears are often amplified by a lack of accessible information or mistrust in corporate and governmental transparency. The social license to operate, therefore, becomes a crucial factor in project success.

To address these challenges, public engagement must be a central component of CCS planning and implementation. Developers and policymakers should prioritize early and continuous dialogue with affected communities, involving them in decision-making processes and providing clear, evidence-based information. Transparency about risks, safety protocols, and long-term monitoring commitments can help build trust and mitigate opposition. Moreover, sharing the economic benefits of CCS projects—such as job creation, infrastructure investment, and community development—can foster a sense of ownership and support among local populations.

Equity, Justice, and Perceived Environmental Trade-offs

The political and social legitimacy of CCS is also tied to broader questions of environmental justice and equity. In some contexts, CCS facilities are proposed in or near disadvantaged communities that already face disproportionate environmental burdens. This raises concerns about whether CCS will contribute to or exacerbate existing inequalities. If the benefits of CCS—such as emissions reductions and technological investments—are not equitably distributed, public resistance and political opposition are likely to intensify.

Addressing these concerns requires that equity considerations be explicitly integrated into CCS policies and project planning. This includes ensuring fair access to funding opportunities, equitable siting practices, and community benefit agreements that compensate for potential risks or disruptions. Incorporating the voices of marginalized groups in consultations and governance structures can enhance legitimacy and reduce opposition. Policymakers should also ensure that CCS is deployed in a way that complements, rather than competes with, renewable energy investment, reinforcing its role in a just and inclusive energy transition.

Media Narratives and the Influence of Civil Society

The media plays a powerful role in shaping public attitudes toward CCS, and negative media coverage can amplify concerns and reinforce political hesitancy. Misreporting of scientific findings, high-profile project failures, or sensationalized safety fears can damage the public image of CCS and undermine its perceived legitimacy. On the other hand, success stories and positive narratives can help build confidence in the technology. Civil society organizations, including environmental NGOs, trade unions, and grassroots groups, are influential actors in the CCS debate. Their support or opposition can sway public opinion and shape political decisions. Engaging with these groups proactively, rather than defensively, can open avenues for constructive dialogue and collaboration. Co-designing CCS projects with input from civil society can enhance transparency, align goals, and build alliances that strengthen both political and social backing.

Addressing Fossil Fuel Industry Involvement and Perceptions of Greenwashing

Another politically sensitive aspect of CCS deployment is its association with the fossil fuel industry. Many major CCS initiatives are led by oil and gas companies, prompting accusations of greenwashing and concerns that the technology is being used to prolong fossil fuel extraction under the guise of climate action. These perceptions can hinder support from climate activists, policymakers, and segments of the public. To counter these criticisms, it is essential that CCS projects are embedded within credible, science-based pathways to net-zero emissions. Governments and regulators must ensure that CCS is not used as a cover for continued fossil fuel expansion, but rather as a tool to decarbonize hard-to-abate sectors and enable negative emissions. Independent oversight, robust environmental criteria, and transparent reporting mechanisms can help distinguish genuine mitigation efforts from attempts to delay the energy transition.

8.4 INNOVATIONS IN CARBON UTILIZATION AND NEGATIVE EMISSIONS TECHNOLOGIES

8.4.1 CO_2-to-Products: Future Markets for Captured Carbon

The transformation of captured CO_2 into marketable products represents an innovative and promising pathway within the broader effort to combat climate change. While not a substitute for deep emission reductions and a transition to renewable energy, CO_2-to-products offers an opportunity to close the carbon loop and add economic value to what was once seen as waste. By creating new markets and revenue streams, it also helps make carbon capture technologies more economically attractive and politically palatable. As research advances and supportive policies take shape, the commercialization of CO_2-based products is likely to accelerate. The key to success lies in continued innovation, cross-sector collaboration, and strong regulatory frameworks that encourage investment and ensure environmental integrity. If developed responsibly and at scale, the CO_2-to-products sector can contribute meaningfully to a sustainable, low-carbon economy, transforming the role of carbon from problem to solution.

CO_2-Based Fuels: A Bridge for Transport and Aviation

One of the most promising markets for captured CO_2 is the production of synthetic fuels, also known as e-fuels. These fuels are produced by combining captured CO_2 with hydrogen derived from renewable electricity, resulting in hydrocarbons that can serve as drop-in replacements for gasoline, diesel, or jet fuel. The use of CO_2-based fuels is particularly important for sectors like aviation and maritime transport, where electrification remains technically and economically challenging. Synthetic fuels made from captured CO_2 offer a way to decarbonize existing fuel infrastructure without the need for massive changes in vehicles or logistics systems. If powered by renewable energy, these fuels can be nearly carbon-neutral over their lifecycle. Countries such as Germany and Norway are already investing in pilot projects and demonstration plants to produce e-kerosene for sustainable aviation fuel (SAF), aligning with international efforts to reduce emissions from air travel. As regulatory frameworks evolve and carbon pricing becomes more stringent, the commercial viability of CO_2-based fuels is expected to improve, supported by mandates, subsidies, and demand from companies seeking to meet low-carbon targets.

Chemical Feedstocks: A Low-Carbon Alternative

Another key application of captured carbon lies in the chemical industry, where CO_2 can be used as a feedstock for producing a variety of chemicals such as methanol, urea, formic acid, and polycarbonates. Traditionally, these chemicals are derived from fossil fuels, contributing significantly to industrial emissions. Substituting CO_2 as a raw material can reduce the carbon intensity of these products and open new pathways for sustainable chemical production.

Methanol synthesis, in particular, has received considerable attention. Methanol is a versatile platform chemical used in fuels, solvents, and plastics. When produced using CO_2 and green hydrogen, it can serve as a low-emission substitute for conventional methanol, which is typically derived from natural gas. Similarly, CO_2-based urea, a critical component of fertilizers, has the potential to reduce emissions in the agricultural sector. These applications are not only technically feasible but also align with existing industrial processes, allowing easier integration into current systems.

To scale such applications, strong market demand, clear certification standards, and supportive policy mechanisms will be required. The chemical industry's interest in reducing Scope 3 emissions, which includes emissions from the supply chain and product use, further strengthens the case for CO_2-derived feedstocks.

Construction Materials: Turning CO_2 into Building Blocks

The construction sector represents another significant market for captured CO_2, particularly through the development of carbon-infused concrete and other building materials. Cement production is one of the largest industrial sources of CO_2 emissions globally, due to both the energy required to produce clinker and the chemical process of calcination. By capturing and mineralizing CO_2, companies can not only reduce emissions but also improve the performance of construction materials. CarbonCure, Solidia Technologies, and other innovative companies have developed methods to inject CO_2 into concrete during the curing process, where it reacts with calcium to form stable carbonates. This process not only sequesters CO_2 permanently but also strengthens the concrete, offering performance benefits alongside environmental ones. Other approaches involve producing aggregates or carbon-negative bricks using CO_2 as a reactant.

Given the massive global demand for building materials, especially in rapidly urbanizing regions, the market potential for CO_2-based construction products is substantial. However, widespread adoption will depend on the development of industry standards, certifications, and procurement policies that recognize the carbon benefits of such products. Governments can play a role by mandating low-carbon materials in public infrastructure projects, thereby creating early markets and accelerating adoption.

Plastics and Polymers: Closing the Carbon Loop

The use of CO_2 in plastics and polymers represents a futuristic yet increasingly feasible market for captured carbon. Researchers and companies are developing processes to convert CO_2 into polyols, which are key building blocks for polyurethane foams used in furniture, insulation, and automotive parts. Unlike traditional petrochemical feedstocks, CO_2-based polyols can significantly lower the carbon footprint of plastic products without compromising performance. By integrating captured CO_2 into consumer goods, companies can offer climate-friendly alternatives that appeal to environmentally conscious consumers. This also opens up marketing opportunities for carbon-negative branding and product differentiation. As awareness of plastic pollution and climate change grows, the demand for low-carbon materials is likely to rise, creating a niche but important market for CO_2-based polymers. Scaling production will require overcoming technical challenges, including reaction efficiency and feedstock purity. However, continued investment in research and development, along with collaboration between academia and industry, will be crucial in unlocking the potential of CO_2-derived plastics. As these technologies mature, regulatory support through eco-labeling and product standards can help accelerate market entry.

Emerging Consumer Markets and Niche Applications

Beyond industrial uses, captured CO_2 is finding its way into niche consumer markets, such as carbonated beverages, specialty chemicals, and even fashion and design. Companies are experimenting with using captured CO_2 to produce synthetic diamonds, textiles, and packaging materials. While these markets are relatively small in terms of volume, they play a valuable role in raising public awareness about carbon capture technologies and demonstrating the creative possibilities of carbon reuse. These applications serve as important proof-of-concept models, showcasing how carbon can be repurposed in everyday products. Their success could influence public perception of CCS and CCU, helping shift the narrative from carbon as a pollutant to carbon as a resource. Moreover, by engaging consumers directly, these markets contribute to broader efforts to mainstream sustainability and foster a carbon-conscious culture.

Economic Viability and Market Drivers

For CO_2-to-products to become commercially viable at scale, several economic and policy factors must align. Chief among them is the cost of CO_2 capture, which remains a major component of the overall production expense. Advances in capture efficiency, modular technologies, and economies of scale will be critical in lowering these costs over time. Additionally, the availability of cheap renewable electricity, particularly for hydrogen production, will greatly influence the competitiveness of CO_2-derived products. Carbon pricing mechanisms, including carbon taxes and emissions trading schemes, can play a central role in leveling the playing field between conventional and CO_2-based products. By internalizing the environmental cost of emissions, these instruments make low-carbon alternatives more attractive. Regulatory mandates, procurement policies, and certification schemes can also create stable demand and provide market signals to investors and developers. Another important consideration is lifecycle assessment and product verification. For CO_2-based products to be recognized as sustainable, clear standards must be established to evaluate their net climate impact. This includes accounting for the energy inputs, the permanence of CO_2 sequestration, and the end-of-life treatment of the product. Transparent and credible certification will be essential in building consumer and investor confidence.

8.4.2 Carbon Capture as a Climate Solutions Portfolio

Carbon capture is not a silver bullet, but it is a critical piece of the climate solutions puzzle. Its unique capabilities—reducing emissions from hard-to-abate sectors, enabling negative emissions, integrating with hydrogen and renewables, and supporting economic transitions—make it indispensable in the global effort to reach net-zero. As climate deadlines approach and the cost of inaction grows, dismissing or delaying carbon capture is a risk the world cannot afford. Inclusion of carbon capture in national climate plans, corporate decarbonization strategies, and international cooperation platforms will be vital for unlocking its full potential. Alongside renewables, energy efficiency, behavioral changes, and nature-based solutions, carbon capture completes the picture of a robust, inclusive, and science-aligned approach to climate mitigation. By treating carbon capture as an essential pillar of a comprehensive climate solutions portfolio, humanity can move more decisively toward a sustainable and stable future.

Science and Mechanism behind Carbon Capture

Carbon capture involves three main components: capturing carbon dioxide, transporting it, and either storing it permanently underground or utilizing it in industrial applications. CO_2 can be captured from large emission sources such as power plants, cement factories, or steel mills through pre-combustion, post-combustion, or oxy-fuel combustion technologies. Once captured, it is compressed and transported via pipelines, ships, or trucks to a storage site. These storage sites are typically deep geological formations, such as depleted oil and gas fields or saline aquifers, where the CO_2 is injected and securely trapped. The efficacy of this process has been demonstrated through several operational projects around the world, including Norway's Sleipner and Snøhvit projects, Canada's Boundary Dam, and the Petra Nova facility in the United States. With appropriate monitoring and risk management, CO_2 storage has been shown to be safe and long-lasting, offering a high-confidence solution for preventing atmospheric emissions.

Addressing Emissions from Hard-to-Decarbonize Sectors

One of the strongest arguments for incorporating carbon capture into a climate solutions portfolio is its unique ability to reduce emissions from sectors that are technically or economically challenging to decarbonize. Industries such as cement, steel, chemicals, and refining inherently produce CO_2 through their industrial processes, not just through energy consumption. For example, in cement production, limestone (calcium carbonate) must be calcined, releasing CO_2 as a byproduct. These process emissions are unavoidable with current methods and cannot be eliminated simply by switching to renewable energy. In such cases, carbon capture becomes the only practical means of achieving significant emissions reductions. Similarly, in heavy-duty transport and aviation, where batteries and hydrogen face limitations in energy density and infrastructure, synthetic fuels made from captured CO_2 offer a viable pathway to low-carbon mobility. As a result, carbon capture is often referred to as an enabling technology for deep decarbonization in the most emissions-intensive parts of the economy.

Enabling Negative Emissions through Direct Air Capture and Bioenergy

Another key contribution of carbon capture in the climate portfolio is its ability to generate negative emissions when combined with certain technologies. Direct Air

Capture (DAC) systems extract CO_2 directly from the atmosphere, which can then be stored underground or used in products. When DAC is paired with geological storage, it becomes a net-negative emissions technology, actively reducing the concentration of atmospheric carbon. Bioenergy with Carbon Capture and Storage (BECCS) is another negative emissions strategy, where biomass is used to generate power or produce fuels, and the resulting CO_2 is captured and stored. Since biomass absorbs CO_2 during growth, capturing and storing the emissions from its combustion results in a net removal of carbon. These negative emission technologies are critical to offset residual emissions from sectors that are unlikely to reach zero, such as agriculture and certain industrial processes. The IPCC and IEA have both emphasized that achieving global net-zero goals without the inclusion of negative emissions technologies would be extremely difficult, if not impossible. Therefore, investing in carbon capture is not only about reducing emissions but also about building the infrastructure necessary for large-scale carbon removal in the coming decades.

Supporting a Just and Inclusive Transition

Carbon capture can also play a role in ensuring a just and inclusive energy transition. Many carbon-intensive industries are located in regions where they provide a significant source of employment and economic activity. Abruptly shutting down these industries without alternatives risks social disruption and economic decline. Carbon capture offers a way to decarbonize these sectors while preserving jobs and enabling regional economic continuity. In regions with a strong fossil fuel legacy, such as the U.S. Gulf Coast, parts of the Middle East, and industrial zones in Asia, carbon capture can be a strategic tool for leveraging existing infrastructure and workforce skills. Repurposing oil and gas pipelines for CO_2 transport, retrofitting power plants, and deploying carbon capture hubs can stimulate investment and create high-value employment. In this way, CCS helps bridge the gap between climate ambition and socio-economic realities, promoting equity in the transition to a low-carbon future.

Integration with Renewable Energy and Hydrogen Systems

Carbon capture does not compete with renewable energy technologies; rather, it complements them in a diversified climate strategy. As intermittent renewable sources like wind and solar continue to scale, flexible and dispatchable generation will remain necessary to balance the grid and ensure reliability. Carbon capture can be applied to fossil-fuel or biomass-fired power plants to provide this reliability while minimizing emissions. Furthermore, carbon capture is closely linked with the development of a low-carbon hydrogen economy. Hydrogen produced via natural gas reforming with CCS, also known as blue hydrogen, can serve as a transitional solution while green hydrogen production from electrolysis scales up. This can help build market demand and infrastructure for hydrogen, accelerating its adoption in transport, industry, and heating. By integrating with renewable systems and enabling low-carbon fuels, carbon capture reinforces the overall resilience and flexibility of the energy system. It supports a smooth decarbonization pathway without over-reliance on any single technology, thus increasing the chances of success in meeting climate targets.

Overcoming Challenges and Accelerating Deployment

Despite its potential, the deployment of carbon capture technologies faces significant challenges. These include high upfront costs, a lack of market incentives, public perception issues, and regulatory uncertainty. Capturing and storing CO_2 requires substantial capital investment, and in many regions, there is no direct financial return unless supported by carbon pricing or government subsidies. Public skepticism also persists, particularly around the long-term safety of geological storage and concerns that CCS might be used to justify continued fossil fuel use. Addressing these concerns requires transparency, rigorous monitoring, community engagement, and robust policy frameworks. To accelerate deployment, governments must provide strong policy signals, such as tax credits, grants, low-interest financing, and emissions performance standards. Collaboration across sectors and borders is equally essential. International cooperation on CO_2 transport infrastructure, knowledge-sharing, and joint research initiatives can lower costs and speed up technological advancement. Scaling up CCS will require coordinated action from industry, academia, civil society, and policymakers.

CHAPTER 9

Case Studies in Carbon Capture and Storage

9.1 CCS IN POWER PLANTS: CHALLENGES AND SUCCESS STORIES

Carbon capture in power plants is a critical but underutilized component of the global climate solutions portfolio. It offers a practical path to reducing emissions from fossil-fuel-based electricity generation while maintaining energy reliability and supporting a just transition. Although significant challenges remain, a growing number of success stories demonstrate that with the right combination of technology, policy, and public engagement, CCS can be effectively deployed in the power sector. As countries strive to meet net-zero targets and navigate the complexities of energy transition, investing in and scaling up CCS will be indispensable to achieving deep decarbonization and long-term climate stability.

9.1.1 Technical and Economic Challenges in CCS Deployment

The integration of CCS into power plants presents a range of technical and economic challenges that have limited its scalability. One of the primary issues is the high cost of capture, which accounts for the largest share of the total CCS project expense. Retrofitting existing plants with capture systems often requires substantial modifications, affecting plant efficiency and increasing operational costs. Power generation with CCS typically experiences an energy penalty, meaning that a portion of the plant's energy output is consumed in capturing and compressing CO_2. This results in reduced net electricity output and higher costs per megawatt-hour of electricity delivered to the grid.

Another major barrier is the lack of infrastructure for CO_2 transport and storage. Developing a full CCS chain requires not only capture equipment at the plant but also pipelines or other transportation systems to deliver CO_2 to suitable storage sites, which may be located hundreds of kilometers away. These sites must be carefully selected and monitored to ensure safe, long-term containment of carbon dioxide in geological formations such as saline aquifers or depleted oil and gas reservoirs. The permitting process for storage sites can be complex, time-consuming, and subject to public resistance. Financing CCS projects in power generation remains another challenge. With the high capital investment required and uncertain long-term returns, many private investors are hesitant to support such projects without strong government incentives. In regions where carbon pricing is weak or absent, there is little economic motivation for utilities to invest in CCS compared to cheaper alternatives like renewable energy or natural gas plants without capture technology. Consequently, policy support, subsidies, and tax credits are critical to making CCS financially viable.

9.1.2 Regulatory and Social Barriers

In addition to technical and financial hurdles, regulatory and social factors also play a role in slowing CCS deployment in the power sector. In many jurisdictions, there is a lack of clear and comprehensive regulatory frameworks governing CCS projects, especially in areas related to liability, long-term monitoring, and storage site closure. Uncertainty around legal responsibilities and environmental compliance can deter investment and delay project timelines. Public perception and opposition also influence CCS development. Some communities express concerns about the safety of underground CO_2 storage, fearing potential leaks or induced seismic activity. Others view CCS as a means to prolong the use of fossil fuels, rather than a genuine step toward climate mitigation. Addressing these concerns requires transparent stakeholder engagement, robust risk assessments, and independent monitoring to build public trust in CCS as a responsible climate solution.

9.1.3 Success Stories: Global Examples of CCS in Power Plants

Despite the challenges, several CCS projects in the power sector have achieved operational success and demonstrate the viability of the technology. These projects offer valuable lessons on how to navigate technical, financial, and policy hurdles while delivering meaningful emissions reductions.

One of the most well-known examples is the Boundary Dam project in Saskatchewan, Canada. Operated by SaskPower, this coal-fired power plant became the world's first commercial-scale power station with CCS when it launched its capture unit in 2014. The system is capable of capturing up to one million tonnes of CO_2 per year, which is either used for enhanced oil recovery or stored underground. Boundary Dam has proven that retrofitting an existing plant with CCS is technically feasible, although the project also highlighted challenges related to cost overruns and operational reliability in the early phases.

In the United States, the Petra Nova project in Texas was another landmark CCS initiative. Commissioned in 2017, Petra Nova captured over 90% of the CO_2 from a 240-megawatt slipstream of a coal-fired unit at the W.A. Parish power plant. The captured CO_2 was transported via pipeline and used in enhanced oil recovery operations. While Petra Nova was temporarily shut down in 2020 due to low oil prices and pandemic-related economic stress, its initial performance showcased the potential for CCS in commercial power operations. The project also benefited from financial support through the U.S. Department of Energy and tax incentives, demonstrating the importance of policy in enabling CCS success.

Another noteworthy development comes from the United Kingdom, where the Drax Power Station is testing bioenergy with carbon capture and storage (BECCS). As part of its transition from coal to biomass, Drax is piloting CCS technology on its biomass units to create a carbon-negative power system. The initiative aims to capture millions of tonnes of CO_2 annually once fully operational, contributing to the UK's net-zero ambitions. Drax's approach illustrates how CCS can be integrated into a broader decarbonization strategy that includes renewable biomass and negative emissions.

9.1.4 The Path Forward: Scaling CCS in the Power Sector

To unlock the full potential of CCS in power generation, coordinated action is needed across policy, industry, and research. Governments must create stable and supportive regulatory environments that incentivize investment and reduce financial risk. This includes implementing carbon pricing mechanisms, offering tax credits or grants, streamlining permitting processes, and supporting CO_2 transport and storage infrastructure. The U.S. 45Q tax credit and the EU Innovation Fund are examples of how targeted financial tools can stimulate CCS deployment.

Continued technological innovation is also essential to improving the cost and performance of capture systems. Research efforts should focus on developing more energy-efficient solvents, membranes, and sorbents, as well as modular and scalable CCS solutions that can be adapted to different plant configurations. Digital monitoring tools and AI-driven predictive models can enhance the safety and efficiency of CO_2 storage, increasing public confidence and operational reliability.

International collaboration will play a key role in knowledge-sharing, standard-setting, and capacity building. Countries with advanced CCS expertise can partner with emerging economies to support project development, transfer technology, and create global networks of CO_2 transport and storage hubs. These efforts can help build the necessary scale to make CCS a mainstream part of global climate strategies.

9.2 CCS IN INDUSTRIAL EMISSIONS: STEEL, CEMENT, AND OTHER HEAVY INDUSTRIES

Industrial activities account for a significant portion of global greenhouse gas emissions, with steel, cement, and other heavy industries being some of the most carbon-intensive. Unlike the power sector, which can shift relatively easily to low-carbon alternatives like renewables, industrial sectors face more complex decarbon-ization pathways. These industries emit CO_2 not only from energy consumption but also from chemical reactions intrinsic to their processes. For example, cement production releases CO_2 during the calcination of limestone, while steel production emits carbon through the reduction of iron ore using coke. These process-related emissions cannot be avoided simply by switching to renewable energy. Therefore, carbon capture and storage (CCS) has emerged as a crucial technology for mitigating emissions in heavy industry, where few alternatives exist. Carbon capture and storage is not merely an option but a necessity for deep decarbonization of the industrial sector. In emissions-intensive industries like steel, cement, and chemicals, where few alternatives exist for reducing process-related CO_2, CCS offers a viable and increasingly proven solution. While cost, infrastructure, and policy challenges persist, a growing number of successful projects demonstrate what is possible with the right combination of technological innovation, financial support, and political will. As the world moves toward a net-zero future, scaling up CCS in heavy industry will be crucial to ensuring that economic growth and climate stability go hand in hand.

9.2.1 Role of CCS in the Steel Industry

The global steel industry contributes approximately 7–9% of global CO_2 emissions, making it one of the top industrial emitters. Traditional blast furnace-basic oxygen furnace (BF-BOF) steelmaking is highly carbon-intensive, relying on coal to reduce iron ore into molten iron. Even though innovations like hydrogen-based direct reduced iron (DRI) methods are being developed, such alternatives remain in their infancy and require considerable investment and supportive infrastructure. As a result, CCS remains one of the most viable near-term solutions to significantly cut emissions from existing steel production facilities.

CCS can be applied in various parts of the steelmaking process, including capturing CO_2 from blast furnace gas or from flue gas streams in power and heat generation units integrated within steel plants. Projects like the LEILAC (Low Emissions Intensity Lime and Cement) initiative in Europe are exploring how CCS can be integrated with steel and cement production. Moreover, the HIsarna project in the Netherlands, backed by Tata Steel and other partners, demonstrates a new steelmaking process that inherently produces a concentrated CO_2 stream, making capture more feasible and cost-effective.

However, retrofitting existing plants with CCS remains a complex engineering challenge. High capital costs, the need for continuous operation, and the integration of capture systems without compromising plant performance are all significant considerations. Nevertheless, with strong policy support and technological advancements, CCS has the potential to become a key driver of decarbonization in the steel sector, particularly during the transition to low-carbon steelmaking.

9.2.2 Decarbonizing Cement Production with CCS

Cement manufacturing is responsible for around 7–8% of global CO_2 emissions, largely due to the calcination process that releases CO_2 when limestone is heated. Unlike emissions from fossil fuel combustion, these process emissions are unavoidable with current methods. This makes CCS not just a complementary option but an essential component for decarbonizing cement. Applying CCS in cement plants typically involves capturing CO_2 from the kiln exhaust gases. This presents specific challenges because cement flue gases have a lower CO_2 concentration compared to those from power plants, and they often contain dust and other particulates that can interfere with capture technologies. Despite these hurdles, several pilot projects are showing promise.

The Norcem plant in Brevik, Norway, is one of the most advanced examples of CCS in the cement sector. With support from the Norwegian government and partners like Heidelberg Materials, this facility is developing a full-scale carbon capture system expected to be operational in the coming years. Once active, it is projected to capture around 400,000 tonnes of CO_2 annually, making it one of the first cement plants in the world to demonstrate commercial-scale CCS. Another initiative is the Carbon Clean project in the UK, which focuses on developing modular, low-cost capture technologies tailored to smaller industrial sites, including cement. These innovations, combined with government funding and industry cooperation, are critical to enabling broader adoption of CCS in the cement industry.

9.2.3 CCS in Other Industrial Sectors

Beyond steel and cement, several other industries emit significant volumes of CO_2 and can benefit from CCS deployment. These include chemical manufacturing, refining, pulp and paper, and waste-to-energy facilities. Many of these industries produce high-purity CO_2 streams, which are ideal candidates for capture due to their lower energy requirements and operational complexity.

In the chemical sector, particularly in ammonia and hydrogen production, CO_2 is generated as a by-product in large concentrations. CCS can be efficiently integrated into these processes, and in some cases, it already has been. For instance, several plants in the United States and Canada capture CO_2 from ammonia production and use it for enhanced oil recovery or store it underground. The Quest project in Alberta, Canada, operated by Shell, captures CO_2 from hydrogen production in an oil sands upgrader and stores it deep underground, showcasing how CCS can be effectively deployed in industrial operations. Refineries also present opportunities for CCS, particularly from hydrogen reformers and fluid catalytic cracking units. However, the integration of CCS in these facilities often depends on site-specific factors such as proximity to storage sites and the ability to modify existing systems without disrupting production. Nevertheless, as demand for low-carbon fuels grows and carbon pricing becomes more widespread, incentives for refineries to adopt CCS are likely to strengthen.

9.2.4 Challenges to Scaling CCS in Industry

While the technical potential of CCS in industrial sectors is clear, its expansion faces several persistent challenges. One of the most significant barriers is cost. Industrial CCS projects often require substantial upfront investment, and many facilities lack the financial resources or incentives to pursue such projects independently. The availability of funding mechanisms, such as government grants, tax incentives, and carbon credits, can therefore make a decisive difference in project feasibility. Infrastructure constraints are another key issue. Industrial sites are often not located near suitable geological storage formations, and building CO_2 transport networks can be expensive and politically sensitive. Coordinated development of shared infrastructure, such as CO_2 pipelines and storage hubs, is essential for reducing costs and enabling regional clusters of CCS deployment.

There is also the challenge of regulatory uncertainty. Many jurisdictions lack comprehensive policies and permitting procedures for CO_2 storage, and questions remain around long-term liability and monitoring responsibilities. These uncertainties can discourage investment and delay project implementation. To overcome them, governments must establish clear and consistent regulatory frameworks that provide legal and financial assurance to project developers. Finally, public perception and social acceptance play a crucial role. Communities may resist CCS projects due to concerns about safety, environmental risks, and the fear that CCS is a distraction from more sustainable solutions. Building public trust through transparent communication, stakeholder engagement, and third-party oversight is vital to gaining support for industrial CCS initiatives.

9.2.5 The Path Forward: Enabling Industrial CCS Deployment

To realize the full potential of CCS in industrial emissions reduction, a multi-faceted strategy is required. First, governments must prioritize industrial CCS in national and international climate policies, recognizing its critical role in achieving net-zero targets. This includes integrating CCS into emissions trading schemes, offering targeted subsidies, and supporting research and development.

Second, industry collaboration is essential. Establishing industrial clusters or hubs where multiple facilities share CO_2 capture, transport, and storage infrastructure can dramatically reduce costs and increase efficiency. Examples like the Net Zero Teesside and Humber Zero projects in the UK show how collaborative models can drive progress. Third, international cooperation on technology transfer, funding, and knowledge sharing is needed to support CCS adoption in developing economies where industrial growth is most rapid. Many of these countries face the dual challenge of economic development and emissions reduction, and CCS can provide a bridge between the two.

Lastly, continued innovation in capture technologies, storage methods, and CO_2 utilization pathways can help reduce costs and expand the scope of CCS applications. From modular capture units to mineralization techniques that turn CO_2 into stable solid products, the technological frontier of CCS is constantly evolving and holds promise for transforming heavy industry into a cleaner, more sustainable sector.

9.3 ROLE OF BECCS IN CLIMATE MITIGATION: CASE EXAMPLES

Bioenergy with Carbon Capture and Storage (BECCS) is increasingly recognized as a pivotal technology in global climate mitigation strategies, particularly for achieving net-zero and net-negative emissions targets. BECCS combines biomass-based energy generation with carbon capture and storage, effectively removing carbon dioxide from the atmosphere. Biomass, derived from plant material, captures atmospheric CO_2 through photosynthesis during its growth. When this biomass is used to produce electricity or fuels and the resulting CO_2 emissions are captured and permanently stored underground, the process results in net-negative emissions. This unique attribute sets BECCS apart from most other mitigation technologies that only reduce emissions without removing them. The Intergovernmental Panel on Climate Change (IPCC) has highlighted BECCS as a key component in scenarios that limit global warming to 1.5°C. However, while its theoretical potential is vast, practical deployment is still limited and surrounded by technical, environmental, and socio-political challenges. Nonetheless, early case studies across different countries offer insights into the viability and future of BECCS.

BECCS has a unique and valuable role in climate mitigation as one of the few technologies capable of delivering negative emissions. Its ability to offset residual emissions from sectors like aviation and industry makes it an essential component of comprehensive climate strategies. Case studies from the United Kingdom, the United States, Sweden, and Japan illustrate how BECCS can be effectively deployed across different sectors and geographies, provided there is sufficient technical capacity, policy support, and public engagement. To realize its potential, BECCS must be scaled responsibly. This includes ensuring sustainable biomass supply chains, reducing costs through innovation, and developing supportive infrastructure and regulation. As countries around the world strive to meet net-zero targets, BECCS can serve as a bridge between current emissions realities and future climate goals. When implemented wisely and collaboratively, BECCS can transform bioenergy systems into powerful tools for atmospheric carbon removal and global climate stabilization.

9.3.1 BECCS as a Negative Emissions Technology

BECCS occupies a distinctive place in the spectrum of carbon reduction solutions because of its ability to generate negative emissions. Most mitigation technologies focus on curbing the flow of new emissions, but BECCS actively removes historical CO_2 from the atmosphere. This makes it especially valuable in offsetting emissions from hard-to-decarbonize sectors such as aviation, shipping, and heavy industry. BECCS is also attractive for countries with significant agricultural and forestry resources, which can supply the biomass feedstock required for energy generation. By integrating with existing biomass power plants or biofuel production facilities, BECCS can convert carbon-neutral processes into carbon-negative ones, enhancing their climate mitigation capacity.

However, the effectiveness of BECCS is highly dependent on the sustainability of biomass sourcing, the efficiency of carbon capture technology, and the availability of secure geological storage sites. If biomass is harvested unsustainably, it can lead to deforestation, land degradation, and food security issues, which undermine the environmental benefits of the technology. Thus, for BECCS to contribute meaningfully to global climate goals, it must be implemented with rigorous sustainability safeguards.

9.3.2 Case Example: Drax Power Station, United Kingdom

One of the most prominent examples of BECCS implementation is found at the Drax Power Station in North Yorkshire, England. Originally a coal-fired power plant, Drax has undergone a significant transformation, converting four of its six generating units to use biomass—primarily wood pellets sourced from sustainably managed forests. In partnership with technology developers and supported by the UK government, Drax has been piloting carbon capture on its biomass units to demonstrate the feasibility of BECCS at scale.

The pilot project at Drax involves capturing CO_2 from biomass combustion and testing various technologies to enhance capture efficiency and reduce costs. The long-term goal is to develop a commercial-scale BECCS facility capable of capturing and storing millions of tonnes of CO_2 annually. If successful, this project could make Drax one of the world's first carbon-negative power plants. The initiative is a cornerstone of the UK's broader net-zero strategy and serves as a testbed for integrating BECCS into a decarbonized electricity grid.

The Drax case also illustrates the critical role of government policy in enabling BECCS. The UK government has included BECCS in its Clean Growth Strategy and provides support through innovation funds and potential carbon pricing mechanisms. This financial and regulatory backing is essential to address the high upfront costs and technical uncertainties associated with BECCS.

9.3.3 Case Example: Illinois Industrial CCS Project, United States

In the United States, the Illinois Industrial CCS project provides another important example of BECCS in action. This project, developed by Archer Daniels Midland (ADM) in Decatur, Illinois, captures CO_2 from a bioethanol production facility. Corn is used as the feedstock to produce ethanol, and during fermentation, nearly pure CO_2 is emitted. This CO_2 is captured and injected into a deep saline formation more than 2,000 meters underground.

The Illinois project has successfully demonstrated the technical feasibility of capturing CO_2 from bio-based industrial processes and storing it securely. It has also shown how BECCS can be integrated into existing industrial operations with minimal disruption. Since 2017, the facility has captured and stored over one million tonnes of CO_2 annually. This case highlights the role of BECCS in the biofuel sector, which is a critical area for decarbonizing transportation fuels.

Moreover, the project has benefited from significant support from the U.S. Department of Energy and state-level environmental agencies. This underscores the importance of public-sector investment and clear regulatory frameworks in driving early-stage BECCS deployment. The Illinois Industrial CCS project is part of a broader national strategy to explore carbon management options and supports regional initiatives like the Midwest Carbon Sequestration Partnership.

9.3.4 Emerging Efforts in Sweden and Japan

Beyond the UK and the U.S., countries like Sweden and Japan are also exploring BECCS as part of their climate strategies. In Sweden, several pulp and paper mills are investigating BECCS technologies to reduce emissions from biomass combustion. Given that these facilities already use sustainable wood residues as fuel, they are well-suited to become early adopters of BECCS. The Swedish Energy Agency is actively supporting feasibility studies and pilot projects to assess how BECCS can be integrated with district heating systems and industrial plants.

In Japan, BECCS is being explored as part of the country's plan to achieve carbon neutrality by 2050. Japan faces unique challenges due to its limited land area and high energy import dependence, but it is investing in innovative BECCS approaches, including algae-based biomass and waste-to-energy processes combined with carbon capture. Pilot projects supported by the government and private sector are helping to build technical expertise and evaluate long-term feasibility.

9.3.5 Challenges and Limitations of BECCS

While BECCS holds promise, its deployment faces several limitations that must be addressed to ensure its responsible and effective use. One of the most debated issues is the availability and sustainability of biomass. Large-scale BECCS could place significant pressure on land, water, and forest resources, potentially competing with food production and biodiversity conservation. To mitigate these risks, strict sustainability standards must govern biomass sourcing, and emphasis should be placed on using agricultural residues, forest waste, and other non-food biomass. Another challenge is the cost. BECCS is currently more expensive than many other mitigation technologies due to the combination of biomass costs, capture infrastructure, and storage development. Without strong policy support, such as carbon pricing or direct subsidies, BECCS may not be economically viable in many regions. Infrastructure requirements also pose barriers. Successful BECCS projects need access to secure geological storage sites and CO_2 transport systems, which may not exist near all biomass or biofuel facilities. Building this infrastructure requires long-term planning, substantial investment, and community engagement. Public acceptance is another hurdle. Misconceptions about biomass harvesting, the safety of CO_2 storage, and the perceived role of BECCS in extending fossil fuel dependency can lead to resistance. Transparent communication, inclusive planning, and demonstrable environmental benefits are essential to build trust and acceptance.

9.4 CO_2-EOR (ENHANCED OIL RECOVERY): EFFICIENCY AND CONTROVERSY

Carbon dioxide-enhanced oil recovery (CO_2-EOR) is a method that uses injected CO_2 to increase the amount of crude oil extracted from an oil reservoir. Traditionally applied in mature or declining oil fields, CO_2-EOR enhances recovery by reducing the viscosity of the remaining oil and improving its flow toward production wells. It represents a long-standing practice within the petroleum industry and has been commercially used since the 1970s, particularly in North America. With the advent of carbon capture and storage (CCS) technologies, CO_2-EOR has gained renewed attention as a possible climate mitigation tool, since the injected CO_2 can potentially be stored underground, reducing atmospheric emissions.

However, the coupling of CO_2-EOR with carbon sequestration raises fundamental questions about its climate value, efficiency, and ethical implications. CO_2-EOR occupies a unique and often polarizing position at the intersection of carbon management and fossil fuel production. Its efficiency as an oil recovery method is well established, and its potential to facilitate early CCS deployment cannot be ignored. However, its credibility as a climate mitigation strategy depends entirely on how it is practiced and governed. If structured correctly—with anthropogenic CO_2, rigorous monitoring, and a clear path toward net emissions reductions—CO_2-EOR can serve as a bridge technology during the energy transition. But if misused, it risks reinforcing the very carbon-intensive systems that the world must leave behind. Therefore, the future of CO_2-EOR must be guided not only by technical efficiency but by an uncompromising commitment to climate integrity.

9.4.1 Technical Efficiency and Operational Benefits

From a technical standpoint, CO_2-EOR is considered one of the most effective tertiary recovery methods in the oil industry. Conventional oil recovery techniques typically leave a large proportion—up to two-thirds—of the original oil in place. CO_2-EOR can improve recovery rates by 10 to 20 percent, depending on reservoir characteristics and the injection method used. This process works best in reservoirs with certain geological properties, such as high permeability and adequate structural traps, which help in both oil recovery and CO_2 containment.

The process generally involves the continuous or alternating injection of CO_2 and water into an oil reservoir. The injected CO_2 dissolves in the crude oil, causing it to swell and reduce its viscosity, thereby increasing its mobility. As the pressure in the reservoir builds, the mobilized oil is pushed toward production wells. Some of the injected CO_2 returns to the surface with the oil and is typically separated and reinjected, creating a closed-loop system that enhances both oil recovery and the potential for long-term CO_2 storage.

One of the practical advantages of CO_2-EOR is its compatibility with existing oil infrastructure. Mature oil fields with declining productivity can be repurposed for CO_2-EOR without the need for entirely new development, lowering capital investment compared to new oil projects. Moreover, the infrastructure used for EOR, such as injection wells, monitoring systems, and pipeline networks, can serve dual purposes for permanent geological storage, creating a bridge between traditional oil production and CCS deployment.

9.4.2 The Role of CO_2-EOR in Carbon Management Strategies

Proponents argue that CO_2-EOR can play a strategic role in early CCS deployment by providing a commercial use for captured CO_2. In this view, EOR acts as a financial catalyst for carbon capture projects, making them more economically viable at current carbon prices. By selling the captured CO_2 for EOR, emitters can offset some of the costs of capture and transportation, helping to stimulate investment in large-scale CCS infrastructure. Additionally, CO_2-EOR can contribute to net carbon reductions when it leads to permanent storage of CO_2 underground. In some advanced EOR operations, more CO_2 is injected than is emitted from the additional oil produced, potentially making the process carbon-neutral or even carbon-negative. This outcome depends on the source of the CO_2 (e.g., industrial vs. natural), the volume of CO_2 injected, the rate of storage retention, and the lifecycle emissions from the produced oil. If all these factors are favorable, CO_2-EOR may offer a transitional climate benefit while more sustainable energy solutions mature.

9.4.3 Case Studies and Real-World Applications

One of the most well-known examples of CO_2-EOR is the Permian Basin in the United

States, where thousands of wells have used CO_2 to recover additional oil over the last few decades. In this region, CO_2 is often sourced from natural underground reservoirs, though there has been a growing interest in using anthropogenic CO_2 captured from industrial sources, such as natural gas processing or ethanol production facilities. Projects like Occidental Petroleum's operations in Texas and the Petra Nova project in Houston illustrate how CO_2-EOR is integrated with large-scale carbon capture. The Weyburn-Midale project in Saskatchewan, Canada, is another key example. Here, CO_2 from a coal gasification plant in North Dakota is piped across the border to Canadian oil fields, where it is injected for EOR. The project is notable for its extensive monitoring and verification protocols, which have provided valuable data on CO_2 behavior in subsurface reservoirs. It is often cited as evidence that CO_2-EOR can function as a form of reliable, long-term carbon storage under the right conditions.

9.4.4 Environmental and Ethical Controversies

Despite its technical effectiveness and commercial benefits, CO_2-EOR remains highly controversial in environmental and climate policy circles. The core of the criticism lies in the fact that CO_2-EOR extends the life of fossil fuel infrastructure and results in the production of additional oil, which, when combusted, releases CO_2 into the atmosphere. Critics argue that promoting oil recovery under the guise of climate mitigation is counterproductive and risks locking in continued fossil fuel dependence.

There is also concern that CO_2-EOR could divert attention and resources away from truly renewable and zero-emission alternatives. Subsidizing EOR projects with public climate funds or carbon credits may undermine efforts to phase out fossil fuels and transition to clean energy systems. The perception that CO_2-EOR serves as a loophole for oil companies to claim environmental credentials while continuing high-emission operations fuels skepticism about its legitimacy as a climate solution.

Another layer of controversy involves the source of the CO_2 used in EOR. Many operations continue to rely on CO_2 extracted from naturally occurring underground reservoirs, which contributes new emissions to the atmosphere rather than mitigating existing ones. Until anthropogenic CO_2 sources are fully prioritized and required, CO_2-EOR will struggle to meet stringent climate integrity standards.

9.4.5 Regulatory and Policy Landscape

The policy treatment of CO_2-EOR varies across jurisdictions and greatly influences its environmental outcomes. In the United States, tax incentives like the 45Q credit provide financial rewards for storing captured CO_2, including through EOR, as long as specific monitoring and reporting standards are met. These credits are designed to encourage the use of anthropogenic CO_2 and ensure long-term storage, but concerns remain about accountability, leakage risks, and whether the emitted oil offsets the stored CO_2. In contrast, the European Union has shown greater caution toward CO_2-EOR, focusing more on permanent geological storage without associated fossil fuel production. Some EU member states prohibit CO_2-EOR outright or impose strict sustainability criteria, reflecting a more skeptical view of its role in a climate-neutral future. This divergence highlights the geopolitical and ideological divides over whether CO_2-EOR should be embraced as a transition tool or avoided due to its fossil fuel associations.

9.4.6 Balancing Efficiency with Long-Term Climate Goals

The central dilemma of CO_2-EOR lies in reconciling its operational efficiency with the ethical imperative of reducing fossil fuel dependence. From a technological perspective, it offers a practical and scalable method for both enhancing oil recovery and sequestering CO_2. In regions where carbon storage options are limited, or where the economic case for CCS is weak, EOR can provide a viable entry point for carbon management. However, for CO_2-EOR to align with long-term climate goals, it must evolve beyond its legacy as an oil production technique and adopt stringent environmental standards. One possible solution is to limit EOR operations to projects that use exclusively anthropogenic CO_2, demonstrate net-negative emissions, and operate within a strict carbon accounting framework. Such projects could be certified under internationally recognized carbon standards and monitored for leakage, lifecycle emissions, and storage permanence. Transparency, regulation, and independent verification are key to transforming CO_2-EOR from a controversial practice into a credible part of the climate solution portfolio.

9.5 DIRECT AIR CAPTURE PROJECTS: COMMERCIALIZATION AND SCALE-UP

Direct Air Capture (DAC) represents one of the most promising yet technologically and economically challenging approaches to removing carbon dioxide directly from the atmosphere. Unlike point-source carbon capture systems that trap emissions at power plants or industrial sites, DAC targets the diffuse CO_2 already present in ambient air. This capability is essential for offsetting emissions from sectors that are difficult to decarbonize, such as aviation and agriculture, and for achieving net-zero or even net-negative emissions in the long term. DAC has gained significant attention in recent years as climate targets have tightened and the urgency for carbon removal has intensified. Direct Air Capture is rapidly transitioning from a futuristic concept to a practical tool in the global effort to combat climate change. While still in its infancy, the technology has proven capable of capturing CO_2 from ambient air and integrating with renewable energy and geological storage systems.

Early commercial projects have demonstrated both feasibility and the challenges that lie ahead. As climate goals become more ambitious, DAC is likely to play an increasingly prominent role in delivering durable and verifiable carbon removal. By fostering innovation, building infrastructure, and aligning market incentives, the world can accelerate the commercialization and scale-up of DAC as part of a comprehensive strategy to achieve climate neutrality and restore the planet's carbon balance.

9.5.1 Principles of Direct Air Capture and Its Distinct Features

The fundamental principle behind DAC is the chemical capture of atmospheric CO_2 using specially designed filters or chemical sorbents. Two main technologies are currently in use: liquid solvent systems and solid sorbent systems. In the liquid approach, air is passed through a chemical solution that selectively reacts with CO_2. In the solid approach, air flows over materials that physically adsorb CO_2 onto their surfaces. Once the CO_2 is captured, it is separated from the sorbent through a regeneration process that typically involves heat, pressure, or both. The purified CO_2 can then be compressed for storage in geological formations or utilized in various industrial applications. DAC is distinct from traditional carbon capture technologies in that it is not tied to a specific emission source. This means it can be deployed nearly anywhere, especially in regions with abundant renewable energy and access to CO_2 storage reservoirs. Additionally, DAC does not require large-scale pipelines or retrofitting of existing infrastructure, giving it unique geographic flexibility. However, these advantages come at a high cost. The low concentration of CO_2 in the air—just over 400 parts per million—makes DAC an energy-intensive process. Overcoming this challenge is essential for commercial viability.

9.5.2 Early Movers and Notable Commercial Projects

Among the pioneers in DAC technology, Climeworks and Carbon Engineering stand out for their progress in moving from laboratory research to commercial-scale projects. Climeworks, based in Switzerland, has developed modular DAC systems using solid sorbents. Its first major commercial project, Orca, was launched in Iceland in 2021 and is capable of removing 4,000 tonnes of CO_2 per year. Powered entirely by renewable geothermal energy, Orca combines DAC with permanent CO_2 storage in basaltic rock formations through a partnership with Carbfix. This project has become a landmark example of how DAC can be integrated into a climate-friendly and scalable carbon removal ecosystem.

Carbon Engineering, based in Canada, employs a liquid solvent approach and has been working on a large-scale DAC plant in the United States in partnership with Occidental Petroleum. The facility, located in Texas, is expected to remove up to 500,000 tonnes of CO_2 annually, with potential for expansion. This project reflects a growing trend of integrating DAC with enhanced oil recovery (EOR) or other utilization strategies to improve the economic return. It also highlights the importance of collaboration between technology developers, energy companies, and government agencies in advancing commercial deployment. Another important player is Global Thermostat, a U.S.-based company focused on developing scalable DAC units using amine-based solid sorbents. While still in the pilot phase, Global Thermostat has emphasized the modularity and cost-competitiveness of its systems, partnering with research institutions and private investors to explore commercial applications.

9.5.3 Scaling Challenges: Cost, Energy, and Infrastructure

Despite encouraging progress, DAC faces substantial challenges that must be addressed to enable large-scale deployment. The most critical barrier is cost. Current estimates place the cost of DAC between $500 and $1,000 per tonne of CO_2 removed, though some optimistic projections suggest future reductions to around $100 to $200 per tonne as the technology matures and scales. Achieving these reductions will require innovation in materials, energy integration, and manufacturing processes.

Energy demand is another central issue. DAC systems require significant thermal and electrical energy for sorbent regeneration and air processing. If powered by fossil fuels, this energy demand could undermine the climate benefits of carbon removal. Therefore, DAC must be paired with low-carbon or renewable energy sources to ensure net-negative outcomes. This limits viable deployment locations to areas with abundant clean energy, which in turn raises concerns about land use, grid integration, and competition with other sectors for renewable power.

Additionally, the infrastructure for transporting and storing captured CO_2 is still underdeveloped in many regions. DAC facilities need reliable access to geological storage sites or industrial customers for CO_2 utilization. Building this infrastructure will take time, policy support, and long-term investment, all of which are still evolving in many parts of the world.

9.5.4 Policy and Market Drivers for Commercialization

The commercialization of DAC depends not only on technological readiness but also on supportive policy frameworks and robust markets for carbon removal. In recent years, there has been a surge in public and private interest in funding DAC projects. Governments in the United States, Canada, the European Union, and the United Kingdom have introduced grant programs, tax credits, and procurement mechanisms to promote DAC. For instance, the U.S. Inflation Reduction Act significantly increased the 45Q tax credit for carbon removal, offering up to $180 per tonne for DAC with permanent storage. This makes DAC projects more financially viable and attractive to investors. Meanwhile, voluntary carbon markets are beginning to recognize the value of durable carbon removal. Companies like Microsoft, Stripe, and Shopify have committed to purchasing high-quality carbon removal credits from DAC providers, helping to create early demand and provide a price signal. These corporate buyers are motivated by science-based climate targets and sustainability commitments, and they often serve as anchor customers for first-of-a-kind projects. Standardization and verification of carbon removal are also essential to build trust in DAC. International frameworks and protocols are being developed to ensure the permanence, additionality, and transparency of DAC-based carbon credits. These efforts are necessary to prevent greenwashing and establish a reliable foundation for a global carbon removal market.

9.5.5 Integration with Broader Climate Solutions

While DAC alone cannot solve the climate crisis, it plays a critical role in complementing other mitigation strategies. It is particularly valuable in offsetting residual emissions from sectors that are unlikely to achieve zero emissions by mid-century. In addition, DAC offers a form of climate insurance, enabling carbon removal to continue even as decarbonization accelerates. Its modular nature allows for flexible deployment, which could prove advantageous in future climate emergencies or efforts to restore atmospheric CO_2 to pre-industrial levels. To maximize its impact, DAC should be integrated with other carbon management solutions, including carbon capture and storage (CCS), nature-based removal strategies, and clean energy expansion. Hybrid systems that combine DAC with biomass energy or industrial processes may offer new efficiencies and cost reductions. For example, co-locating DAC facilities with renewable energy installations or waste heat sources can reduce operational costs and emissions.

9.5.6 The Path Forward: From Demonstration to Deployment

The path from demonstration to large-scale deployment of DAC is challenging but increasingly navigable. The lessons learned from early projects like Orca and the Texas-based Carbon Engineering facility will shape the next generation of DAC systems. As more pilot plants come online and as technology providers gain experience, economies of scale and learning-by-doing are expected to drive down costs and improve performance. Governments, industries, and civil society must work together to overcome the barriers to commercialization. This includes sustained public investment in research and development, targeted incentives for early deployment, robust carbon markets, and international collaboration. Public awareness and engagement will also be essential, as DAC projects may raise concerns about land use, energy consumption, and corporate accountability.

CHAPTER

Appendices

10

10.1 GLOSSARY OF KEY TERMS IN CCS

- Basalt Formations: Volcanic rocks that can react chemically with CO_2 to form solid carbonate minerals, offering an alternative long-term storage option through mineralization.
- BECCS (Bioenergy with Carbon Capture and Storage): A carbon-negative process where biomass is used to produce energy and the resulting CO_2 emissions are captured and stored, removing CO_2 from the atmosphere.
- Caprock: An impermeable layer of rock that sits above a storage reservoir, preventing CO_2 from migrating to the surface. A secure caprock is essential for effective geological storage.
- Capture: The first step in the CCS process, in which CO_2 is separated from gases produced in electricity generation and industrial processes. Capture technologies include pre-combustion, post-combustion, and oxy-fuel combustion methods.
- Carbon Capture and Storage (CCS): A climate change mitigation technology that captures carbon dioxide emissions from sources like power plants or industrial facilities and stores it underground in geological formations to prevent it from entering the atmosphere.
- Carbon Credit: A tradable certificate representing the removal or reduction of one tonne of CO_2 or its equivalent from the atmosphere. CCS projects may generate credits if properly verified.
- Carbon Dioxide (CO_2): A colorless, odorless greenhouse gas that is emitted through natural processes like respiration and volcanic eruptions, and through human activities such as burning fossil fuels. It is the primary gas targeted for capture in CCS.
- Carbon Intensity: The amount of CO_2 emitted per unit of energy or product. CCS aims to lower the carbon intensity of electricity, fuel, and industrial materials.

- Carbon Lock-In: A situation where investment in fossil fuel infrastructure makes it harder to transition to low-carbon alternatives. CCS can be both a solution and a contributor to lock-in if not managed carefully.
- Carbon Market: A system where carbon credits can be bought and sold to incentivize emission reductions. CCS can participate in these markets if projects meet regulatory standards.
- Carbon Neutral: A state where CO_2 emissions are balanced by carbon removal or offset activities. CCS can help achieve this by capturing and storing unavoidable emissions.
- Carbon Removal: Processes that actively remove CO_2 from the atmosphere, such as Direct Air Capture or Bioenergy with Carbon Capture and Storage (BECCS), and store it to reduce overall atmospheric levels.
- Carbon Utilization: The conversion of captured CO_2 into valuable products, such as building materials, fuels, or chemicals. Sometimes referred to as CCUS (Carbon Capture, Utilization, and Storage).
- Compression: After capture, CO_2 is compressed into a dense, fluid-like state to allow for efficient transport and injection. This high-pressure form is known as supercritical CO_2.
- Depleted Oil and Gas Fields: Former hydrocarbon reservoirs that have been emptied of their oil and gas. They are considered suitable for CO_2 storage because they have already demonstrated the ability to contain fluids for millions of years.
- Direct Air Capture (DAC): A technology that captures CO_2 directly from ambient air rather than from a concentrated source like a factory or power plant.
- Enhanced Oil Recovery (EOR): A technique where CO_2 is injected into aging oil fields to extract additional oil. While this process results in some CO_2 storage, its climate benefits are debated due to the additional fossil fuel extraction.
- Flue Gas: The mixture of gases emitted from power plant chimneys or industrial stacks. It is the primary target for post-combustion CO_2 capture.
- Geological Storage: A form of CO_2 storage that involves injecting the gas into porous rock formations deep underground. The rock traps the gas and prevents it from escaping to the surface.

- Greenhouse Gas (GHG): A gas that traps heat in the atmosphere and contributes to global warming. CO_2 is the most significant anthropogenic GHG, making its capture and storage vital.
- Hard-to-Decarbonize Sectors: Industries such as cement, steel, and aviation that are difficult to transition away from fossil fuels due to technical and economic constraints. CCS is considered essential for reducing emissions in these areas.
- Leakage: The unintended escape of stored CO_2 back into the atmosphere or surrounding environments. Preventing leakage is critical to the credibility and safety of CCS.
- Life Cycle Assessment (LCA): An analysis of the environmental impact of a technology or product throughout its entire life cycle. For CCS, this includes capture, transport, storage, and monitoring.
- Mineralization: A natural or engineered process where CO_2 reacts with certain minerals (like calcium or magnesium) to form stable carbonates, effectively locking the CO_2 in solid form.
- Monitoring, Verification, and Accounting (MVA): A system used to ensure that stored CO_2 remains underground and does not leak. MVA includes tools and procedures for tracking CO_2 movement and verifying storage integrity.
- Net-Zero Emissions: A state in which the total amount of greenhouse gases emitted is balanced by an equivalent amount of emissions removed or offset, often requiring technologies like CCS and DAC.
- Oxy-Fuel Combustion: A technique that burns fossil fuels in oxygen instead of air, resulting in a flue gas that is mostly CO_2 and water vapor, which makes CO_2 easier to capture.
- Permanence: The requirement that stored CO_2 remain securely trapped for long periods, typically thousands of years, to effectively mitigate climate change.
- Porosity and Permeability: Key rock properties that determine how well CO_2 can be injected and retained. High porosity allows for storage volume, while appropriate permeability facilitates CO_2 movement during injection.
- Post-Combustion Capture: A process in which CO_2 is removed from flue gases after fossil fuels have been burned. It is the most widely used and can be retrofitted to existing plants.

- Pre-Combustion Capture: A method where CO_2 is removed before fossil fuels are burned. This is done by converting fuel into a mixture of hydrogen and CO_2, separating the CO_2, and using the hydrogen for power generation.
- Public Acceptance: The level of support or opposition from communities and stakeholders toward CCS projects. Social trust, transparency, and local engagement are critical for project success.
- Reservoir Integrity: The ability of a geological formation to contain CO_2 over the long term without significant leakage or migration. Ensuring integrity is crucial for storage safety.
- Saline Aquifer: A deep underground porous rock formation filled with salty water (brine) that can store CO_2. These are among the most promising storage options due to their large capacity and global availability.
- Storage (Sequestration): The process of injecting compressed CO_2 into deep underground rock formations, such as depleted oil and gas fields or deep saline aquifers, where it can be securely stored for centuries or longer.
- Supercritical CO_2: A state where CO_2 behaves like both a gas and a liquid, occurring at high pressure and temperature. This form is ideal for transport and storage due to its high density.
- Transport: The movement of captured CO_2 from the point of capture to a storage site. This is typically done using pipelines, though ships and trucks may also be used.

10.2 RESOURCES AND RECOMMENDED READING

As a rapidly evolving field, CCS requires continuous engagement with both foundational materials and emerging research. The resources listed above provide a multifaceted view of CCS technologies and policies, catering to a diverse audience of engineers, policymakers, investors, students, and environmental advocates. Whether one is seeking a basic understanding or an expert-level analysis, these readings and platforms serve as gateways to mastering the complexities and opportunities within Carbon Capture and Storage. As global efforts intensify to combat climate change, staying informed about CCS will be essential to supporting its responsible and effective deployment.

10.2.1 Intergovernmental Panel on Climate Change (IPCC) Special Reports

One of the most authoritative sources for climate science, the IPCC's special reports include detailed analyses of CCS. The 2005 IPCC Special Report on Carbon Dioxide Capture and Storage remains a foundational document, providing comprehensive coverage of the science, technologies, costs, and risks of CCS. More recent assessments, including the IPCC's Sixth Assessment Report (AR6), also integrate CCS as part of the broader climate mitigation strategies, especially in scenarios that aim to reach net-zero emissions.

10.2.2 International Energy Agency (IEA) Publications

The IEA publishes regular updates and roadmaps on CCS development and deployment. Key titles include "Carbon Capture, Utilisation and Storage: A Vital Clean Energy Technology" and "CCUS in Clean Energy Transitions." These documents offer data-driven insights into global CCS trends, project development, regulatory environments, and technology readiness. The IEA's role as an impartial observer of international energy markets makes its reports essential for understanding CCS in the context of energy transitions.

10.2.3 Global CCS Institute Reports and Database

The Global CCS Institute is a leading organization focused entirely on advancing CCS technologies. It maintains an up-to-date Global Status of CCS Report, which is published annually and tracks the progress of CCS projects around the world. This report includes data on commercial, pilot, and demonstration projects, as well as detailed commentary on policy developments, market trends, and investment flows. The Institute's online library and database are excellent resources for finding case studies and regional analyses.

10.2.4 U.S. Department of Energy (DOE) and NETL Resources

The U.S. DOE and its National Energy Technology Laboratory (NETL) have been at the forefront of CCS research, especially in the context of power generation and industrial decarbonization. The DOE's Carbon Capture Program provides technical reports, funding opportunities, and roadmaps for future development. NETL offers modeling tools, technology baseline reports, and cost assessments that are particularly valuable for engineers and project developers looking to evaluate technical feasibility and economic performance.

10.2.5 European Commission and Research Programmes

In Europe, the European Commission has funded CCS initiatives through its Horizon 2020 and Horizon Europe research programs. The Strategic Energy Technology (SET) Plan includes a focus on CCS and CCUS technologies, with emphasis on large-scale demonstration and integration into industrial clusters. Reports from projects such as STRATEGY CCUS and ALIGN-CCUS provide insights into region-specific challenges, regulatory frameworks, and public acceptance issues across EU member states.

10.2.6 Scientific Journals and Academic Literature

Peer-reviewed journals such as International Journal of Greenhouse Gas Control, Energy Procedia, Environmental Science & Technology, and Applied Energy publish cutting-edge research on various aspects of CCS. Topics include capture technology innovations, storage site characterization, environmental impacts, life cycle assessments, and integration with renewable energy systems. Access to academic databases like ScienceDirect, Springer, Wiley, and Taylor & Francis is crucial for readers seeking advanced technical knowledge.

10.2.7 Books for In-Depth Learning

Several books offer structured, comprehensive learning paths for students, researchers, and professionals. Carbon Capture by Howard Herzog provides a readable yet technically informed introduction to the subject, explaining the evolution of CCS and its place in climate policy. Carbon Capture and Storage: Emerging Legal and Regulatory Issues by Ian Havercroft, Richard Macrory, and Richard B. Stewart discusses the complex legal and institutional frameworks necessary for successful CCS implementation.

10.2.8 Project Websites and Company Case Studies

Websites of leading CCS project developers and operators are valuable for understanding real-world applications. For example, the Sleipner Project in Norway and Boundary Dam in Canada provide data on capture rates, storage methodologies, and operational performance. Companies like Climeworks, Carbon Engineering, and Occidental Petroleum publish updates on Direct Air Capture and CCS deployment. These sources often include technical documentation, performance summaries, and sustainability reports that shed light on business models and stakeholder engagement.

10.2.9 Think Tanks and NGOs

Organizations such as the Clean Air Task Force, Bellona Foundation, Carbon180, and the Center for Climate and Energy Solutions (C2ES) produce policy briefs, white papers, and market analysis focused on CCS. Their work often bridges the gap between technical findings and policy implementation, offering guidance on how to align CCS with just transition goals, environmental justice, and economic development. These groups also engage in advocacy and provide recommendations for national and international climate policies.

10.2.10 Educational Videos and Online Courses

For those preferring visual learning, platforms like Coursera, edX, and YouTube host lectures, animations, and short courses on CCS. Universities such as MIT, Stanford, and the University of Edinburgh have created online content that covers the fundamentals of carbon capture technology, storage geology, and climate mitigation strategies. These resources are especially useful for students and non-specialists seeking accessible explanations of complex technical subjects.

10.2.11 Government Climate Plans and Roadmaps

National governments are increasingly incorporating CCS into their long-term climate strategies. The UK's Net Zero Strategy, the U.S. Long-Term Strategy under the Paris Agreement, and Australia's Technology Investment Roadmap all outline roles for CCS in decarbonizing energy and industry. These documents provide context-specific plans, targets, and policy mechanisms that support CCS deployment and often reference the importance of public-private partnerships and infrastructure development.

10.3 KEY ORGANIZATIONS AND CCS RESEARCH INSTITUTES

10.3.1 Global CCS Institute

The Global CCS Institute, headquartered in Australia with offices in several continents, is one of the most influential organizations solely dedicated to accelerating the deployment of CCS worldwide. It acts as a central hub for information sharing, capacity building, and stakeholder engagement. The Institute publishes the annual "Global Status of CCS" report, which tracks the progress of projects across various sectors and regions. It also provides technical analyses, policy briefings, and a digital knowledge hub accessible to governments, companies, and civil society.

10.3.2 International Energy Agency (IEA)

The International Energy Agency plays a pivotal role in assessing global energy trends and technology pathways, including CCS. The IEA's publications such as the "Net Zero by 2050" roadmap and "CCUS in Clean Energy Transitions" outline the significance of CCS in achieving climate goals. The agency supports member and non-member countries through policy advice, scenario modeling, and strategic roadmaps, emphasizing the importance of CCS for decarbonizing heavy industries and ensuring energy security.

10.3.3 Intergovernmental Panel on Climate Change (IPCC)

While not a research institute in the traditional sense, the IPCC provides authoritative scientific assessments that include CCS as a necessary part of many mitigation scenarios. In both its special reports and comprehensive assessments, the IPCC evaluates the effectiveness, costs, and potential of CCS technologies. Its role is critical in legitimizing CCS within the broader climate policy discourse and informing international negotiations under the United Nations Framework Convention on Climate Change (UNFCCC).

10.3.4 National Energy Technology Laboratory (NETL), USA

The U.S. Department of Energy's National Energy Technology Laboratory is a leader in advancing fossil energy research, including CCS technologies. NETL conducts in-house research and funds external projects related to capture materials, storage site characterization, and carbon utilization. It has played a major role in projects like Petra Nova and has contributed to the development of CO_2 pipelines and monitoring technologies. NETL's Carbon Storage Program and associated research partnerships have generated some of the most detailed cost assessments and risk evaluations in the field.

10.3.5 Bellona Foundation

Based in Norway, the Bellona Foundation is an environmental NGO that strongly supports CCS as a necessary tool to combat climate change. It advocates for policy frameworks, industrial decarbonization, and deployment of CCS in Europe. Bellona is actively involved in stakeholder dialogues and contributes to shaping European Union directives on emissions reduction. The foundation's expertise bridges the gap between environmental advocacy and technical CCS advancement.

10.3.6 CO_2CRC (Cooperative Research Centre for Greenhouse Gas Technologies), Australia

CO_2CRC is one of the world's leading collaborative research organizations focused on CCS. It operates the Otway Project in Victoria, which serves as a test site for cutting-edge monitoring and storage technologies. CO_2CRC brings together universities, government agencies, and private industry to conduct applied research on geosequestration, risk management, and community engagement. It serves as a model for cooperative innovation and public-private partnership in CCS development.

10.3.7 Norwegian CCS Research Centre (NCCS)

Located in Norway and coordinated by SINTEF, the NCCS is one of Europe's most active research centers dedicated to CCS. It is involved in several high-profile projects, including the Northern Lights CO_2 transport and storage project and research related to the Sleipner and Snøhvit sites. NCCS focuses on developing cost-effective, integrated CCS systems and aims to establish Norway as a global leader in full-chain CCS deployment.

10.3.8 Carbon Management Canada (CMC)

Carbon Management Canada is a national not-for-profit research organization based in Alberta that works on technologies for CO_2 monitoring, storage validation, and methane detection. It supports field testing through the Containment and

Monitoring Institute (CaMI), which operates a research wellfield for CO_2 injection experiments. CMC's focus on emissions management and subsurface monitoring makes it an essential contributor to Canada's CCS research ecosystem.

10.3.9 UK Carbon Capture and Storage Research Centre (UKCCSRC)

Funded by the UK Research Councils, the UKCCSRC is a network of academic researchers dedicated to advancing CCS technologies. It provides funding, training, and infrastructure for research in capture materials, storage geology, systems integration, and public acceptance. UKCCSRC also plays a central role in linking academia with industry and government stakeholders in the United Kingdom's low-carbon transition strategy.

10.3.10 GCCSI Asia and Middle East Initiatives

The Global CCS Institute, through its regional offices, also supports CCS development in Asia and the Middle East. In countries like China, Japan, and the United Arab Emirates, CCS has gained traction in both industrial and power sectors. Institutes such as China's Administrative Centre for China's Agenda 21 and Saudi Aramco's R&D centers contribute to region-specific CCS adaptation and technology demonstration, ensuring that emerging economies can participate in climate mitigation through CCS.

10.3.11 European Zero Emissions Platform (ZEP)

ZEP is a European technology platform that brings together stakeholders from industry, research, and academia to advocate for CCS deployment in Europe. It works closely with the European Commission to influence policy design, infrastructure planning, and funding allocation for low-carbon technologies. ZEP emphasizes the integration of CCS into industrial clusters and hydrogen production, helping shape Europe's vision for net-zero industry.

10.3.12 Clean Air Task Force (CATF)

This U.S.-based environmental group champions innovation in climate technologies, including CCS. CATF conducts policy research, public education, and technical advocacy to support the deployment of carbon capture in the U.S. and globally. The organization's focus on pragmatic, technology-neutral climate strategies has positioned it as a credible voice among both environmental advocates and industry players.

10.3.13 IEAGHG (IEA Greenhouse Gas R&D Programme)

Operating under the auspices of the International Energy Agency, the IEAGHG conducts specialized research on greenhouse gas mitigation technologies. It organizes technical studies, workshops, and international collaborations focused on storage integrity, capture efficiency, and life cycle analysis. IEAGHG's rigorous approach to scientific inquiry and its convening power make it a central player in international CCS dialogue.

10.4 FREQUENTLY ASKED QUESTIONS (FAQS)

- What is Carbon Capture and Storage (CCS)?
- How does CCS work in practice?
- Why is CCS important in the fight against climate change?
- Is CCS safe for the environment and public health?
- How much does CCS cost, and who pays for it?
- Can CCS be applied to all types of industries?
- What is the difference between CCS and CCUS?
- Where is captured CO_2 stored, and how is it monitored?
- Are there any examples of successful CCS projects?
- What challenges does CCS face today?

Reference

- Advances in Carbon Capture and Storage Technology – Ashok Kumar Gupta, 2022, Elsevier
- Advances in CO_2 Capture – Angelo Basile & Fausto Gallucci, 2021, Elsevier
- Carbon Capture – Jennifer Wilcox, 2012, Springer
- Carbon Capture and Sequestration – Elizabeth Wilson & David Gerard, 2007, Wiley
- Carbon Capture and Storage – Stephen A. Rackley, 2nd ed., 2017, Elsevier
- Carbon Capture and Storage and the Uncertain Future of Clean Coal – Jonatan Pinkse, 2017, Routledge
- Carbon Capture and Storage for Sustainable Energy Future – Malti Goel, 2014, Springer
- Carbon Capture and Storage in Developing Countries – Natalia Kulichenko & Eleanor Ereira, 2012, World Bank Publications
- Carbon Capture and Storage in Energy and Climate Policy – Sara Budinis, 2017, Palgrave Macmillan
- Carbon Capture and Storage in International Energy Policy and Law – D. Russell & D. Reiner, 2011, Oxford University Press
- Carbon Capture and Storage Technologies – Mai Bui & Niall Mac Dowell, 2018, Springer
- Carbon Capture and Storage: Emerging Legal and Regulatory Issues – Ian Havercroft et al., 2nd ed., 2018, Hart Publishing
- Carbon Capture and Storage: Issues and Prospects – Bruce E. Dorn, 2016, Nova Science Publishers
- Carbon Capture and Storage: Opportunities and Challenges for Africa – Njavwa Sumbwanyambe, 2021, IGI Global
- Carbon Capture and Storage: Options for Underground Storage – Adrian H. M. Verweij, 2016, CRC Press

- Carbon Capture and Storage: Physical, Chemical, and Biological Methods – Le Quéré C., 2009, Springer
- Carbon Capture and Storage: Planning and Development – Deborah Gordon, 2015, Wiley
- Carbon Capture and Storage: Research, Development and Demonstration at the Global Scale – Michael B. McElroy, 2014, Nova Science Publishers
- Carbon Capture and Storage: The Solution for Deep Emissions Reduction – Winston R. Alcorn, 2019, Nova Science
- Carbon Capture and Storage: Theoretical Perspectives and Practical Realities – Philip Gass, 2022, Routledge
- Carbon Capture by Chemical Absorption – Sven J. Norin, 2019, Springer
- Carbon Capture for Storage in Deep Geologic Formations (Vol. 1 & 2) – King, Chidambaram & Herzog, 2005, Elsevier
- Carbon Capture, Storage and Utilization: A Possible Climate Change Solution for Energy Industry – Nikolaos Koukouzas, 2021, Springer
- Carbon Dioxide Capture and Storage: A Key Abatement Option – International Energy Agency (IEA), 2008, IEA Publications
- Carbon Dioxide Recovery and Utilization – Mikio Makino & M. Aresta, 2003, Springer
- Carbon Dioxide Sequestration in Geological Media – M. Grobe, 2010, AAPG Studies
- Carbon Dioxide Storage in Oil and Gas Fields – S. M. A. Hosseini, 2021, Elsevier
- Carbon Dioxide Utilization: Electrochemical Conversion into Fuels – Inamuddin & Mohd Imran Ahamed, 2020, Springer
- Carbon Dioxide Utilization: From Fundamental Science to Industrial Applications – Peter Styring, 2015, Elsevier
- Carbon Management Technologies: Current Status and Future Outlook – E. Rubio, 2020, Elsevier
- Carbon Management: Implications for R&D in the Chemical Sciences and Technology – National Research Council, 2001, National Academies Press
- Carbon Sequestration and Its Role in the Global Carbon Cycle – Brian J. McPherson & Eric T. Sundquist, 2009, American Geophysical Union

- Carbon Sequestration in Urban Ecosystems – Rattan Lal & David J. Augustine, 2011, Springer
- Carbon Storage in Sub-Seafloor Geological Structures – Takuo Ishida, 2018, Springer
- CCS Technology in the Energy Sector – David C. Hittle, 2021, Elsevier
- Chemical Looping Systems for Fossil Energy Conversions – Liang-Shih Fan, 2011, Wiley
- CO_2 Capture and Storage: Including Conversion Pathways – Claudia R. V. Morgado, 2020, CRC Press
- CO_2 Emissions and Carbon Capture: A Global Perspective – Syed Tauseef Mohyuddin, 2022, Springer
- CO_2 Geological Storage: Geomechanics and Monitoring – Stéphanie Vialle, 2018, Wiley
- CO_2 Injection and Storage in Deep Saline Formations – Behnam Zakeri, 2020, Elsevier
- CO_2 Injection in Reservoirs for Enhanced Oil Recovery – Ramona Do Nascimento, 2019, Elsevier
- CO_2 Storage Capacity Estimation in Deep Saline Aquifers – Birkholzer, 2013, Lawrence Berkeley National Laboratory
- CO_2 Storage in Carbonate Reservoirs – C. D. White, 2021, Elsevier
- Engineering Aspects of Carbon Capture and Sequestration – Rodney C. Ewing, 2019, American Chemical Society
- Geologic Carbon Sequestration – Mark D. Zoback & Steven M. Gorelick, 2012, Cambridge University Press
- Geological Storage of Carbon Dioxide – Stéphanie Vialle et al., 2018, Wiley
- Integrated Assessment of Carbon Capture and Storage – Selma L'Orange Seigo, 2015, Springer
- Sustainable Carbon Capture – Rahul S. Pophali, 2023, Springer
- The Economics of Carbon Capture and Storage – Vanessa Wallace, 2014, Routledge
- Understanding Carbon Sequestration: Science, Technology, and Policy – Rebecca I. Green, 2021, CRC Press

Index

www.ingramcontent.com/pod-product-compliance
Lightning Source LLC
LaVergne TN
LVHW081938200926
848484LV00009B/2467